Springer Theses

Recognizing Outstanding Ph.D. Research

Aims and Scope

The series "Springer Theses" brings together a selection of the very best Ph.D. theses from around the world and across the physical sciences. Nominated and endorsed by two recognized specialists, each published volume has been selected for its scientific excellence and the high impact of its contents for the pertinent field of research. For greater accessibility to non-specialists, the published versions include an extended introduction, as well as a foreword by the student's supervisor explaining the special relevance of the work for the field. As a whole, the series will provide a valuable resource both for newcomers to the research fields described, and for other scientists seeking detailed background information on special questions. Finally, it provides an accredited documentation of the valuable contributions made by today's younger generation of scientists.

Theses may be nominated for publication in this series by heads of department at internationally leading universities or institutes and should fulfill all of the following criteria

- They must be written in good English.
- The topic should fall within the confines of Chemistry, Physics, Earth Sciences, Engineering and related interdisciplinary fields such as Materials, Nanoscience, Chemical Engineering, Complex Systems and Biophysics.
- The work reported in the thesis must represent a significant scientific advance.
- If the thesis includes previously published material, permission to reproduce this must be gained from the respective copyright holder (a maximum 30% of the thesis should be a verbatim reproduction from the author's previous publications).
- They must have been examined and passed during the 12 months prior to nomination.
- Each thesis should include a foreword by the supervisor outlining the significance of its content.
- The theses should have a clearly defined structure including an introduction accessible to new PhD students and scientists not expert in the relevant field.

Indexed by zbMATH.

Fatemeh Mokhtari

Self-Powered Smart Fabrics for Wearable Technologies

Doctoral Thesis accepted by
University of Wollongong, Australia

Springer

Author
Dr. Fatemeh Mokhtari
Australian Institue for Innovative Materials
University of Wollongong
Wollongong, NSW, Australia

Supervisor
Dr. Javad Foroughi
ARC DECRA Fellow
University of Wollongong
Wollongong, Australia

Visiting Professor
University of Duisburg-Essen
Duisburg, Germany

ISSN 2190-5053 ISSN 2190-5061 (electronic)
Springer Theses
ISBN 978-3-031-06480-7 ISBN 978-3-031-06481-4 (eBook)
https://doi.org/10.1007/978-3-031-06481-4

This Springer imprint is published by the registered company Springer Nature Switzerland AG
The registered company address is: Gewerbestrasse 11, 6330 Cham, Switzerland

This book is dedicated to my dearest husband and my parents, for their endless encouragement, love and support.

Supervisor's Foreword

Since mechanical movement is ubiquitous in real environments, the conversion of ambient mechanical energy to electrical energy using piezoelectric materials appears to be one of the more economic and environmentally friendly useful methods for powering sensors and portable electronic devices. However, current piezoelectric generators face challenges with respect to flexibility, wearability, durability, and industrial fabrication. To address these issues, fiber-based piezoelectric generators have attracted attention due the combination of their unique properties including lightweight and soft nature, as well as natural microstructure. Hence, wearable technology has attracted intensive interest owning to their ability to convert the energy from body movement and the surrounding environment to electronic signals which can power portable/wearable electronic devices for sensing, motion tracking or monitoring to improve the quality of human life.

This book presents the development of innovative methodology to fabricate nanostructured piezoelectric composite fibers mainly for use in wearable technologies application as an energy generator and/or sensors. The thesis considers methods of piezoelectric fiber formation and develops novel textile structures (weave, knit, braid, coil) with embedded electrodes. The flexibility and small diameter of the final fiber make it possible to use them in garment without affecting structure of comfort. The fabricated structures have maximum output voltage of 380 mV and a power density of 29.62 μW cm^{-3} which is ~1559% higher than previously reported for piezoelectric textiles.

Since during COVID-19 pandemic, detection of virus suspected patient through wearable sensors is becoming timely important, and remotely monitoring physiological signals, such as respiration, and heart rate using wearable technology improves clinical decision-making path. Nanostructured piezoelectric fibers presented in this thesis are a novel method with ability of flexibility, lightweight and mass production capability that have great potential application as motion detectors and self-powered biomedical and wearable sensor applications in daily life.

The book targets a wide readership including materials scientists, electrical engineering, soft robotics, Internet of things, electronic textiles and wearable technology. I wish to acknowledge the Dr. Fatemeh Mokhtari for her diligence and effort in bringing the material to the leadership. Hopefully, our readership finds this book timely and informative.

Wollongong, Australia
October 2021

Dr. Javad Foroughi

Abstract

Recent development in miniaturize electronic devices with higher computational capabilities and ultra-low-power communication technologies involves a tendency toward powering these devices with high-energy efficiency, long cycle life, fast and cheap manufacturing and low weight power sources. Mechanical energy harvesters are needed for such diverse applications such as self-powered wireless sensors, structural and human health monitoring systems and cheaply harvesting energy from human movements. Integration of piezoelectric materials and novel fabrication strategies with conventional textile process established the emergence of wearable technology field which can meet these needs. In particular, fiber-based electronic devices can be integrated into garments with desirable attributes such as flexibility, stretchability, permeability and lightweight. Piezoelectric fibers are an ideal interface platform option between environment, electronic devices and human's body, and polyvinylidene fluoride (PVDF) is highlighted in the wearable technology field due to its good chemical resistance, strength, thermal resistance and stable and large piezoelectric, ferroelectric and pyroelectric properties. Previous work has shown that PVDF fibers can be integrated into textile-based mechanical energy harvesters and have highlighted the potential for further improvement. The current challenges include slow processing, low-energy conversion efficiency and difficulty in integrating the fibers (and associated interconnecting electrodes) into textile garments. Therefore, the aim of this project is to fabricate and characterize piezoelectric fibers suitable for ready integration into textile materials. The book considers methods of fiber formation (including the use of additives to enhance the piezoelectric coefficient) and develops novel textile structures (weave, knit, braid, coil) with embedded electrodes. The flexibility and small diameter of the final fiber make it possible to use them in garment without affecting structure or comfort. Finally, the performance of the fiber generators was evaluated through different applications such as air and water sensor, health and movement monitoring and energy generator.

This book is summarized in the sections as follows:

Chapter 1 describes the definition of E-textile, smart textile and wearable technology. The chapter critically reviews the recent development in fiber-based self-powered systems and sensors with special focus on piezoelectric PVDF and composite structures. The principle of piezoelectricity and their charge generation mechanism are considered. In addition, several strategies that may improve the performance of piezoelectric generators are summarized. Furthermore, a summary of the market for wearable technology is considered along with the demands for high-end functional textile such as washability and degradability.

Chapter 2 provides an overall insight on the general experimental fabrication methods as well as the characterization analyses.

Chapter 3 demonstrates novel triaxial braided PVDF yarn harvesters that use the piezoelectric effect to convert tensile mechanical energy into electrical energy. Compressing or bending braided PVDF yarns generated maximum output voltage of 380 mV and power density of 29.62 μWcm^{-3} which is ~1559% higher than previously reported for the piezoelectric textiles. The developed triaxial energy generator exhibits significantly higher sensitivity by a factor of 4 when compared with the PVDF energy generator. Unlike for other piezoelectric harvesters, the triaxial braided PVDF yarn produces tensile energy harvesting with durability which is enabling cycles up to 50% strain for thousands of cycles with no changes in its performance. The production processes are compatible with industrial, large-scale textile manufacturing that can be used for a variety of potential applications such as wearable electronic systems and energy harvesters charged from the everyday body movement.

In Chap. 4, a novel approach is reported to create wearable energy generators and sensors using nanostructured hybrid piezoelectric fibers and exploiting the enormous variety of textile architectures (knitting, braiding and weaving). It is found that high-performance hybrid piezofiber was obtained using a barium titanium oxide (BT) nanoparticle and PVDF with a mass ratio of 1:10. These fibers were knitted to form a wearable energy generator that produced a maximum output voltage of 4 V and a power density 87 μWcm^{-3} which is 45 times higher than earlier report for piezoelectric textiles. The wearable energy generator charged a 10 μF capacitor in 20 s which is 4 and 6 times faster than previously reported for PVDF/BT and PVDF energy generators, respectively. It also emerged that the established knitted energy harvester exhibits sensitivity of 6.3 times higher in comparison with the piezo fibers energy generator. A knee sleeve prototype based on a PVDF/BT wearable device for monitoring real-time precise health care was demonstrated. The developed processing method is scalable for the fabrication of industrial quantities of strain sensing and energy generation smart textiles.

Chapter 5 describes the characterization and synthesis of PVDF nanocomposites fibers and coiled fibers that include a filler mixture of BT nanoparticles and reduced graphene oxide (rGO) nanosheet. Based on FTIR and XRD analysis, there was a 58% improvement in β phase formation for PVDF/rGO/BT nanocomposite coil in comparison with fiber. The dielectric measurements of PVDF/rGO nanocomposites fiber revealed a low dielectric loss factor and high dielectric constant. Young's modulus and tensile strength of the as-prepared PVDF/rGO nanocomposite fiber are 108% and 63% higher than pure PVDF fiber, respectively. Polarization versus electric field (P–E) loop revealed that on an external electric field of 300 kV cm^{-1}, the PVDF/rGO/BT nanocomposite has an energy density of ~80 mJ cm^{-3}, which proves its potential applications for energy storage too. The PVDF/rGO/BT coils can stretch up to ~100% strain and caused a peak voltage output of ~1.3 V with a peak power density of 3 W kg^{-1} which is 2.5 times higher in comparison with previous research. The coil structure shows an energy conversion efficiency of 22.5%. The fabricated coils have great potential for applications in high-performance, flexible, lightweight and stretchable piezoelectric devices for energy harvesting and biomedical applications. The coils were employed in wearable and adjustable bending sensors for human fingers to demonstrate their feasibility in motion capture applications and coils in a woven structure for muscle contraction sensing for athletes.

Chapter 6 provides a summary of the thesis and highlights directions for future research. This thesis demonstrates the feasibility of the piezoelectric fiber production in combination with different fillers and investigated their incorporation into textile structures in a diverse range of applications. Significantly, the approaches shown are relatively simple to fabricate, cheap, flexible and easily incorporated into wearable devices to provide in situ monitoring and energy harvesting.

Parts of this thesis have been published in the following documents:

Journals

1. Mokhtari F., Foroughi J., Zheng T., Cheng Z., Spinks G. M., "Triaxial braided piezo fiber energy harvesters for self-powered wearable technologies", *Journal of Material Chemistry A,* 2019**,** *7* (14), 8245.
2. Mokhtari F., Cheng Z., Raad R., Xic J., Foroughi J., "Piezofiber to Smart Textile: Review on Recent Advances and Future Outlook for Wearable Technology", *J. Mater. Chem. A*, 2020,8, 9496–9522.
3. Mokhtari F., Spinks G. M., Fay C., Cheng Z., Raad R., Xic J., Foroughi J., "Wearable Electronic Textiles from Nanostructured Piezoelectric Fibers", *Advanced Materials Technologies* 5.4 (2020): 1900900.
4. Mokhtari F., Shamshirsaz M., Latifi M., Foroughi J., "Nanofibers-Based Piezoelectric Energy Harvester for Self-Powered Wearable Technologies", *Polymers* 2020, *12*(11), 2697.
5. Mokhtari F., Spinks G. M., Sayyar S., Cheng Z., Ruhparwar A., Foroughi J., "Highly Stretchable Self-Powered Wearable Electrical Energy Generator and Sensors", *Adv. Mater. Technol.* 2021, 6, 2000841.
6. Mokhtari F., Spinks G. M., Sayyar S., Foroughi J., "Dynamic Mechanical and Creep Behaviour of Meltspun PVDF Nanocomposite Fibers", *Nanomaterials* 2021, 11(8), 2153.

International Conferences

1. Mokhtari, F.; Foroughi, J.; Spinks, G. M.; Cheng, Z.; Wallace, G.G., "High-performance melt-spun piezoelectric fiber for self-powered textiles", *ACES International Electromaterials Science Symposium*, Wollongong, Australia, February 2017
2. Mokhtari, F.; Foroughi, J.; Spinks, G. M.; Cheng, Z.; Wallace, G.G., "High-performance piezoelectric fiber for wearable technologies", *Biomimetic, Artificial Muscles and Nano-bio (BAMN) conference*, Wollongong, Australia, September 2017
3. Mokhtari, F.; Foroughi, J.; Spinks, G. M.; Cheng, Z., "Nanostructured hybrid piezoelectric film as self-powered smart textile", *The International Conference on Nanoscience and Nanotechnology*, Wollongong, Australia, February 2018
4. Mokhtari, F.; Foroughi, J.; Spinks, G. M.; Cheng, Z., "Triaxial piezoelectric nanocomposite fibers for energy harvesting application", *6th International Conference on Multifunctional, Hybrid and Nanomaterials*, Sitges, Spain, March 2019

Book Chapters

1. Mokhtari F., Foroughi J., Latifi M., "Enhancing β crystal phase content in electrospun PVDF nanofibers", In book "Energy Harvesting Properties of Electrospun Nanofibers", *IOP Publishing*, 2019; pp 5-1-5-28.

Acknowledgments

I would like to express my sincere appreciation to my Primary Supervisor, Dr. Javad Foroughi, and Co-supervisors, Prof. Geoffrey M. Spinks and Prof. Zhenxiang Cheng, for their excellent supervision, financial assistance and constant encouragement throughout my Ph.D. study. I would also like to thank Mr. Tony Romeo and Dr. Mitchell Nancarrow at Electron Microscopy Centre (EMC) for their help and guidance on SEM and optical microscopes. I would also like to acknowledge the general technical support from Associate Professor Germanas Peleckis, Dr. Vitor Sencadas and Dr. Cormac Fay from university of Wollongong. I appreciate the assistance of Dr. Sepidar Sayyar and Dr. Patricia Hayes on material characterizations. I am grateful to Mr. Adam Taylor and Mr. Ali Jeirani for their guidance in development of 3D designs and laser cutting.

I acknowledge the use of facilities within the Australian National Fabrication Facility (ANFF). A special thanks to the Intelligent Polymer Research Institute (IPRI) for providing a prestigious research environment and funding support through the Australian Research Council (ARC) under Discovery Early Career Researcher Award (DECRA) and University of Wollongong Global Challenges program. I am also appreciative of all staff member and students at the Intelligent Polymer Research Institute (IPRI) who have made these past four years a pleasurable experience.

I want to thank my wonderful family for their constant support and for their many sacrifices that allowed me to be where I am today. I thank my mother Shahrzad, my father Ahmad, my brother Ehsan, for giving me the inspiration and balance I needed to continue. I also want to thank my family in Iran who are so close despite being so far away. This thesis is dedicated to my parents. Finally, I would like to thank my best friend and husband Hassan, without his love and support none of this work would have been possible.

Contents

Abbreviations

3D	Three dimensional
AC	Alternating current
AFM	Atomic force microscopy
BT	Barium titanium oxide
CCG	Chemically converted graphene
D	Diameter
d_{33}	Longitudinal piezoelectric coefficient
DC	Direct current
DMA	Dynamic mechanical analysis
DMF	Dimethylformamide
DSC	Differential scanning calorimetry
E	Electric field
Eq	Equation
F	Force
FTIR	Fourier transform infrared spectroscopy
g	Gram
GO	Graphene oxide
hr	Hour
Hz	Hertz (frequency)
i	Current
L	Liter
mA	milliampere
min	Minute(s)
mL	Milliliter
mV	Millivolt
OLED	Organic light-emitting diode
P	Pressure
PFM	Piezoelectric force microscopy
PVDF	Polyvinylidene fluoride
PZT	Lead zirconate titanate
R	Resistance

rGO	Reduced graphene oxide
rpm	Revolutions per minute
s	Second
SEM	Scanning electron microscope
SS-PFM	Switching spectroscopy piezoresponse force microscopy
T	Temperature
t	Time
Tan δ	Dielectric loss
TGA	Thermogravimetric analysis
wt.%	Weight percent
XRD	X-ray diffraction
Y	Young's modulus
ε	Strain
ε_0	Vacuum permittivity
ε_r	Dielectric constant
°C	Degree of Celsius

Chapter 1
Introduction and Literature Review

1.1 Introduction

The industrialisation approaches create new incorporative fields such as Internet of Things (IoTs) [1], robots, artificial intelligence (AI) and big data [2]. The IoT is a system of interrelated computing devices which comprises everyday objects, such as medical technology, mobile phones, laptops, and other electronic devices, and perhaps one day, it will even include ordinary things such as clothing and furniture [3]. Electronic textiles (e-textile) are poised to play an important role in the IoT development to connect larger number of objects to each other and to the cloud [4]. The technological advances of these devices has revolutionized our world and daily lives with substantial growth in usage expected in the near future. The power sources for these devices are normally rechargeable batteries. These power source have the limitation such as periodic charging times, minimal life-cycle, potential hazard, high-priced disposal and significant environmental risks [5] so they may not be the most appropriate power source for future electronic devices. Therefore, many efforts are underway to find renewable, environmentally friendly and green energy source to power up electronic devices [6–8]. There are various green energy sources including solar [9], wind [10], acoustic and radio wave [11, 12], heat [13, 14], and vibrations [15]. These renewable energy harvesting systems attract attention in the research area of renewable and sustainable energy harvesters due to the rapid spread of wireless electronics [16]. Most of the devices developed during the past decade just had a single source for energy harvesting. However, recently rising interest is in multi-source hybrid energy harvesting [17].

Energy harvesting from body movement and surrounding environment can potentially fulfil such a goal due to ubiquitous availability. The human body movement is an enormous storehouse of energy. Human body has abundant energy from the mechanical movements of jumping, walking, and running and organ motions like respiration and heartbeat [18]. Harvesting from large scale body movements with high mechanical energy, particularly at joints like the knee and elbow, involve complex deformations of the devices in multimodal straining modes such as bending, twisting,

F. Mokhtari, *Self-Powered Smart Fabrics for Wearable Technologies*, Springer Theses,
https://doi.org/10.1007/978-3-031-06481-4_1

pressing, and multiaxial stretching. At the same time, they require not only flexibility but also the stretchability, which are both essential for efficient energy conversion [19]. Body movement as a power source can be considered as an attractive alternative over traditional rechargeable batteries to supply electrical power to low-energy devices such as wearable electronics and body worn sensors [20]. The power from a 68 kg adult's daily activities such as heating, breathing, blood circulatory, arm motion, walking and typing could reach to over 100 W. Converting 1% of this power may be sufficient to charge portable electronics [21].

Clothing is one of the necessities of human life which is considered as the second skin and is the only wearable that adjusts to our everyday lifestyle over the course of a lifetime. Textiles represent a strong candidate to become the next interface between the real and the digital world [22]. Textiles combine useful properties of light weight, soft nature, natural micro-structure, wearable convenience, and air permeability and can be an ideal candidate for wearable applications [23]. The emergence of wearable devices could potentially improve the quality of human life owing to their applications for healthcare monitoring [24], smart fabrics [25], motion tracking, roll-up displays [26], and wearable heaters [27]. Although the first generation of such smart textile only uses the textile as a substrate for rigid and miniaturized electronic components, intrinsically flexible and wearable sensors directly use fabric as functional materials to a seamless integration of the multi-function sensors and textiles [23].

There are three different research categories which attract considerable attention in the area of wearable energy harvesting. First, applying energy harvesting materials in different forms such as fibers or yarn [28], two [29, 30] and three-dimensional [31, 32] fabric and film structures [33]. These provide different performances in the case of flexibility, stretchability and appropriate integration of active materials and electrodes. Second, researchers are considering using polymers and diverse range of fillers which enhance output power, higher energy conversion efficiency and sensitivity of e-textiles [34, 35]. Third, the integration of wearable energy-harvesting with energy-storage devices which is an attractive route that create new opportunities for empowering wearable electronics sustainably [36]. These devices have lighter weight, longer operational time or less frequent recharging which bring convenience of consumers [37].

Since the 1990s the field of energy harvesting has been studied, and the technological progress in this field has accelerated, especially in the last few years [38]. Many devices such as electromagnetic [39], piezoelectric [40], and triboelectric [41] energy harvesting devices have been exhibited to convert body movement energy into electricity, which can be used to power various wearable and implantable electronics [18]. Each energy resource has its own advantages in particular environments. Among them, the piezoelectric energy harvesting technology has particularly attracted attention because of its high energy density characteristics [42]. Piezoelectric materials convert mechanical energy into electricity through a mechanical-to-electrical transformation mechanism [43].

Piezoelectric materials simply can be divided into four classification: single crystals, ceramics, polymers, and composites (polymer/ceramic combination). Among them, piezo ceramics generally suffer from brittle and rigid properties that fails to satisfy the design of flexible wearable electronic devices [44]. Over years, many attempts have been made to achieve high performance of piezoelectric generators, which is generally affected by both intrinsic (crystal structure [45], piezoelectric coefficient [46], geometry effect [45], arrangement [47], etc.) and extrinsic (frequency [48], applied pressure [49], poling effect [50], external load, etc.) parameters [51]. Ferroelectric polymer materials such as poly (vinylidene fluoride) (PVDF) and its copolymers are well known for their piezoelectric properties. Despite they have lower piezoelectric coefficients than their inorganic counterparts, they are flexible and can tolerate the higher mechanical strain [52, 53].

Polar β-phase in PVDF is an essential parameter for efficient energy-harvesting applications [54]. Addition of functional nanomaterials (TiO_2 [55], Fe-rGO [56], $BiVO_4$ [57]) into the PVDF matrix is a novel approach that interrupts the PVDF molecular symmetry and leads to the formation of polar crystalline β-phase and consequently enhance the ferroelectric properties. Also, the various nanomaterials with inherently high piezoelectric properties (PZT [58], BT [59], $NaNbO_3$ [60]) has been added into the PVDF matrix to increase the overall piezo response [61].

PVDF offer a wide diversity of fabrication methods including, solvent-casting [62], melt-casting [63], fiber extrusion [64], electrospinning [65] and phase-inversion [66]. However, one of the main challenges for all these materials and methods is to transform them in proper textile structures to sustain the basic properties of the fabrics, such as flexibility and wearability. The electrospun fibers are aligned along length axes and there is no need for poling process due to the high electric field applied through fabricating but still the processes is quite tedious and expensive [65, 67]. The flexible film based materials devices can be bent in one direction. But their lack of low damage tolerance, air-permeability, and poor stretchability limit their application, specifically for wearable application [21]. Therefore, piezoelectric fiber based textile generators are in high demands. Fiber-based flexible generators can be integrated in textile products such as shoe insoles, trousers and clothing designed to harvest energy from daily activity [68].

Early research about textiles generators was performed with a simple woven structure that can transfer force or tensile stresses onto piezoelectric yarns [69]. Recently, textile designs, such as knitted [70], braided [71] coiled structures [72] and 3D printed fabrics [73] have also been used to fabricate wearable piezoelectric textiles.

The development of piezoelectric fibers or yarn leads to the development of piezoelectric textiles. Like most of the polymers, the one-dimensional nature of piezoelectric fibers providing tailorable mechanical properties and chain alignment, presenting piezoelectric properties and a high degree of crystallinity [74]. The energy units (fiber/fabric) for smart garments must be able to cut in different length/shapes to meet versatile integration requirements. Hence, tailorability is an essential property not only for fiber fabrication to textile but also for textile being designed into clothes with different sizes and styles [75]. These considerations have been explored for fiber based triboelectric generators [76–79] but less work has been done in the case of piezoelectric fiber based generators [71, 80, 81].

This chapter summarises the basic classifications and working principles of piezoelectric systems. Recent developments in material selections, fabrication techniques and structural design are considered for wearable technology. In particular, the recent accomplishments in the ferroelectric polymer field, with special focus on PVDF polymer and its composites is discussed along with the potential application and continuing growth in the development of sustainable power source from wearable technology. A brief collection of the fiber-based self-powered devices are introduced and future directions, challenges, opportunities and guideline for future development in the area of the fiber based piezoelectric generators are discussed.

1.2 E-Textile Technologies

Computers and smart phones changed our daily life due to transferring enormous amount of data which connect us to our surroundings. These electronic devices have been developed and now come in different shapes, having in common that they are portable, collecting data smartly and bringing communication and computing much closer to the human body. These fundamental needs have led to a technological convergence growth, most significantly, mechanical suitability, integrating electronics, and communicative ability, which are now being extrapolated to the concept of wearable electronics [82, 83]. Electronic textiles have been considered owing to a raise of internet-of-things (IoTs) and wearable electronics [84]. Electronic Textiles (e-textile) are able to perform electronic functions and are perceived as a way to add features into ordinary wearable textiles and building competitive advantages in market [85, 86]. Since a couple of years, wearable electronics (apparatuses worn directly on the skin in different parts of the body), including shirts, necklaces, shoes, headbands, smart wristbands, watches and eyeglasses have gained considerable attention, because the data collection both continuously and noninvasively in real-time can be easily provided [87]. E-textiles are likely to be built from a wide range of functional devices that may include displays, sensors, microprocessors, battery, communication units, etc. In this case these devices may have quite a different appearance from conventional rigid electronic devices, forming stretchable, flexible, patchable to the skin, and even body implantable. In terms of these aspects, e-textiles seem to be highly appropriate. Textiles benefited from properties including flexibility, light weight, compressibility, and wearability have been evolved into the research and development which covering materials science, electronic technologies and mechanical engineering and making e-textile technology (Fig. 1.1) [76]. In past various electronic devices such as transistors, displays, sensors, and batteries should be integrated to the textile which they were bulky, rigid, heavy and load the users [88]. Today these electronic devices have been developed and help users combining the features of both fashion item and information technology product [89]. Hence, materials science, functional textiles, and apparel design integrated, to lead the rapid growth of new types of fibers, fabrics, and garments [90]. E-textiles attract attention as the next-generation electronics due to their wide range of applications from wearable human-interactive interfaces, biomedical/wellness monitors, to consumer electronics and shape-adaptive military [91]. There were many efforts during the last

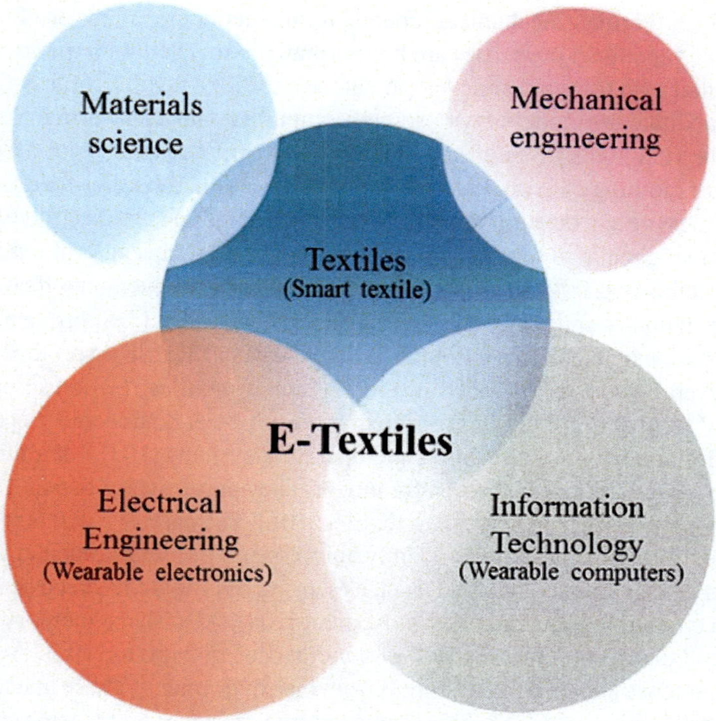

Fig. 1.1 Multidisciplinary environment for e-textiles

decades to design and develop a series of e-textiles with the purpose of being smart and flexible enough to be use in everyday life [92]. These efforts offer a research platform in the area of smart textile and wearable technology. Smart textile becomes a major segment of e-textiles along with smart clothing and information science which will be discussed in detail in the following section.

1.2.1 Smart Textile

Textiles have been at the centre of human technological innovation for thousands of years [93]. Previous textiles consider just one of the necessity of human beings and now has developed out as an innovative area that is able to satisfy human desire to its supreme extent. The textile which focused on preliminary necessities call as "first generation textile" [94]. Nowadays the term "Smart Textiles" refers to a broad research fields and products that develop the usefulness and functionality of common textiles. Smart textiles are defined as textile products such as fibers filaments and yarns in the form of knitted, woven or non-woven structures, which can interact with the environment and user [95]. In a more clear definition, smart textile are described as textiles that can sense and react to environmental conditions or exciter, from thermal, mechanical, chemical, magnetic, electrical, or other sources [96]. These functional properties are not normally associated with traditional textile [97]. Parallel with technology development the need for a flexible information framework to facilitate the model of widespread computing, and a platform for monitoring the human vital signs in an efficient and cost-effective manner with a "universal" interface of clothing essentially led to the birth of the smart textiles field [98]. In fact smart textiles are at a meeting point between electronics and textiles [99] which has a very promising realm in science and technology nowadays because of public interest and marketability [100]. Smart textiles have been used in several applications in the healthcare industry, military, and as wearable electronics [101]. Sporting and fashion industries are expected to contribute more to the smart textile than other industries [102]. Several steps are included in creating smart textiles: (a) fabric production, especially for mass production, (b) integration of power source and electronic part into textiles and (c) communication and operating systems [103]. Smart textiles can sense and react to change in their environment. They may be classified into three categories: passive, active and very smart textiles [101]. The passive smart textiles are the first generation of smart textiles and only able to sense the environmental stimuli and have been used in biomedical field, such as sensing of biophonic, electroencephalography, electrocardiogram and body movement [104]. Also shape memory materials, hydrophilic or hydrophobic textiles, etc. are part of this group [105]. Active smart textiles can sense and respond to stimuli from the environment. These materials use as sensors and actuators [96]. The third category are very smart textiles that consists of a unit for sensing, reasoning and actuating. They are able to detect various signals from environment (light intensity, temperature, pollution, etc.), choose how to respond, and finally to act by using various flexible textile based actuators (textile displays, micro vibrating devices, OLED, LED, etc.) [100]. Totally this type of textiles sense, react and adapt themselves to the environmental conditions [95]. Corresponding to the development of wearable applications, the use of smart textiles increased for real-time monitoring of bioelectrical signals, where a conductive fiber or fabric can be used as dry electrode. Currently, most of the commercial smart textiles are advanced based on this type of sensors [106]. Smart textile item consist of: (a) sensing part, (b) actuating part, (c) communication unit, (d) powering/generating/storing component, (e) data analysis element, and (f) interconnects [99]. In these fields, smart wearable structures provide some advantage such as conformability to human body at rest and in motion, comfort in close contact to skin, and suitability as substrate for smart components [107].

1.2.2 Wearable Technology

"Wearable technology" and "wearable devices" are terms that describe computers and electronics that are integrated into clothing and other accessories that can be worn comfortably on the body. Over recent years, wearable devices have evolved from bulky, large, and uncomfortable technologies to more fashionable, lightweight, and portable items [108]. The ideal intention of wearable technologies is to integrate functional portable electronics and computer seamlessly into people's daily lives. In wearable technology the conductive areas can be established typically by manually attaching conventional wires or sewing conductive yarns [109], printing rigid and stretchable conduction lines [110], replacing non-conductive fibers with conductive ones [111] and other developed technology. The advancement of conducting fibers with desired properties is an important step in the application of fibrous materials in wearable technology. The two most important properties of wearable devices are the breathability and flexibility that provide the fabric-based comfort properties [112]. Embedding microelectronic devices into wearable products like fibers, yarns or fabrics, has the advantage of utilization of the existing strength industries, such as textiles and footwear, jewellery, apparel and electronics [113]. Wearable technology with persuasive features of being breathable, flexible, lightweight, shape conformal and even washable provides a novel way of designing functional next-generation wearable electronics [114]. A critical component in the area of wearable technology specifically for future portable and wearable electronics is fiber electronics which concern electronic devices on textile fibers. Its final applications include the fields of entertainment, finance, education, music and gaming. However, in the fields of health care, medicine, and fitness wearable technology potentially has its greatest impact [115]. Textiles are the main interfacing mechanisms between the naked body and surrounding environment [116] hence they can collect signals from the human body as inputs and provide sufficient response for the monitored person through a connected device [117]. Wearable technologies can be used to manage personal healthcare by early detection of health disorders such as sleep monitoring [118, 119], heart rate [120], blood pressure [121] and breath volatile organic compound sensors [122] to obtain real-time information about performance [123]. The real-time generators are required to achieve long term cycling and efficient electrical output to support continuous operation in environmental monitoring, defence, and healthcare biomedical device applications [124]. The reason of the measuring real time data refer to users' desire to quantify their ability in collecting more information which advances in sensors provide deeper measuring capability. Users learn more about themselves, thus changing in lifestyle can be under their control [89]. Wearable textiles in sportswear could help to enhance athletic personal comfort, performance and protection. They also provide useful information on the athletes' physical abilities, training status, and responses to exercise [26]. it has been proved that the highest production of biomechanical power is formed in the lower limb, at the ankle during the foot-strike [125]. Power consumption, Power source, and energy efficiency are important factors which should be considered in the selection and design of wearable technologies. Wearable sensors have to be extremely power efficient, because frequent battery changes for multiple wearable sensors would likely prevent users' acceptance and increase the final cost [126]. All of the wearable e-textiles have battery demands that must be fulfilled either by detachable or by thin, flat, and flexible batteries that are able to endure washing, drying, ironing, and dry cleaning [85]. To power wearable electronics, commercial rechargeable batteries have been widely used, which require periodic recharging due to their limitation in energy storage capacity. However, repeated battery charging interrupts the continuous monitoring since the devices typically have to be taken off through recharging [127]. Apart from the danger of explosion and possible leakage of electrolyte solution, batteries may cause severe potential accidents to both human body and ambient environment. Therefore, harvesting energy from body movement and ambient environment which can extend batteries life or replace them is an extremely desirable to work with wearable devices always and everywhere [5]. Since body movement is an infinite and chargeless energy source, and clothes are an essential requirement in daily life, textile based generator will become an extremely attractive topic for harvesting body movement energy which is being less affected by the ambient environment [2]. The human body joint points are in a continuous state of motion in a fixed axis which provide an input mechanical source to harvest this motion energy [128]. The criteria for wearable electronic devices are that they must be invisible and comfortable; provide reasonable long reactivity lifetime with proper power output and low-cost fabrication process [81].

Depending on the body motion or signal that require to be harvest or sense, the selection of material is very important in corporation of wearable and flexible electronics. Future predictions for materials selection shows that the materials available in nature for wearable electrodes will get a higher market share, followed by the

Fig. 1.2 Schematic illustration and application demonstration of fiber based piezoelectric generators for and wearable energy harvesters [4, 20, 68, 129–137]

composites made from natural/synthetic sources [102]. Hence piezoelectric materials is proving to be one of the right choice for energy harvesting from body movements [28]. Since the most specific feature of the mechanical movement of the human body is its low frequency nature and the large displacements associated with it [138], the harvester should be highly, soft and comfortable. Many flexible energy harvesters have been reported using various conversion basis, including piezoelectric [34], electrostatic [139], and triboelectric [140] mechanisms. The piezoelectric mechanism is superior to the other conversion mechanisms due to its high energy density and adaptable across a wide range of dimensions [135]. Piezoelectric textiles have been recommended for use in wearable devices due to their stable energy conversion properties, stable mechanical properties and high sensitivities [141]. Power generation could be attained through piezoelectric elements that generate energy from various sources of motion or vibration [95]. Wearable piezoelectric generators can be used for generating power from different human activation including finger, knee or elbow bending and foot stepping to power up portable and wearable electronic device [132]. Figure 1.2 demonstrate diverse applications of textile-based piezo generators on human body.

1.3 Piezoelectric Generators

A variety of ambient energy sources such as thermal, solar, acoustic energy, acoustic noise and vibration have been studied as an additional energy provider for the recent years. Among them, thermoelectric generators, vibration/kinetic driven power generator, and solar cells are more widely studied because of their ubiquity, high efficiency, and potentials to miniaturization [142, 143]. Based on material, structure formation, and working mechanisms, generators can be in three main groups of pyroelectric, triboelectric and piezoelectric [2]. Pyroelectric generators can convert periodically convert thermal energy variation into electricity. However, thermal energy is hard to control, cannot be used for medical implant, and the conversion efficiency is lower comparing with vibration and solar energy sources [13]. Triboelectric [144] and piezoelectric [145] generators enable to efficiently convert mechanical energy to electricity [146]. Recently triboelectric generators have been used to harvest mechanical energy from friction, vibration and even acoustic waves [68]. With similar device size, piezoelectric generators can generate more electric charge than their triboelectric counterparts, so are more suitable to make wearable devices [147]. The literature review provides that the environmental mechanical energy density can reach 1–10 mW cm^{-2}, which has the second place after solar energy (10–100 mW cm^{-2}) [148]. Piezoelectric generators transform mechanical energy to electricity through deformation of piezoelectric material sandwiched between two electrodes, which induces a piezoelectric potential [149]. This feature enables piezoelectric electronics to be easily miniaturized and made/integrated into complex shapes/structures [150]. Whereas human body is a mechanical energy source and also user of wearable electronics, providing self-dependent way for energy utilization is in high demand [2].

Piezoelectric generators attracted researchers and engineers attention in the area of wearable electronics, with nearly 5000 publications in the last decade [150]. Numerous approaches have been used to fabricate piezoelectric generators, such as spinning, coating, printing, and depositing (Fig. 1.3). Film forming technique such as coating, blading, casting and painting are simple and useful technique but their ability may be limited by the dimension, structure, or surface aspect of part [151]. Printing technology provides a favourable solution for mass production and fabrication of less complex shapes. This technique is a high efficiency method to reduce material consumption and final cost [152]. Spinning process such as wet spinning [153], melt spinning [71], dry spinning [154], gel spinning [155], and electrospinning [156] is the common method for fabricating flexible piezoelectric generator with the ability for mass production.

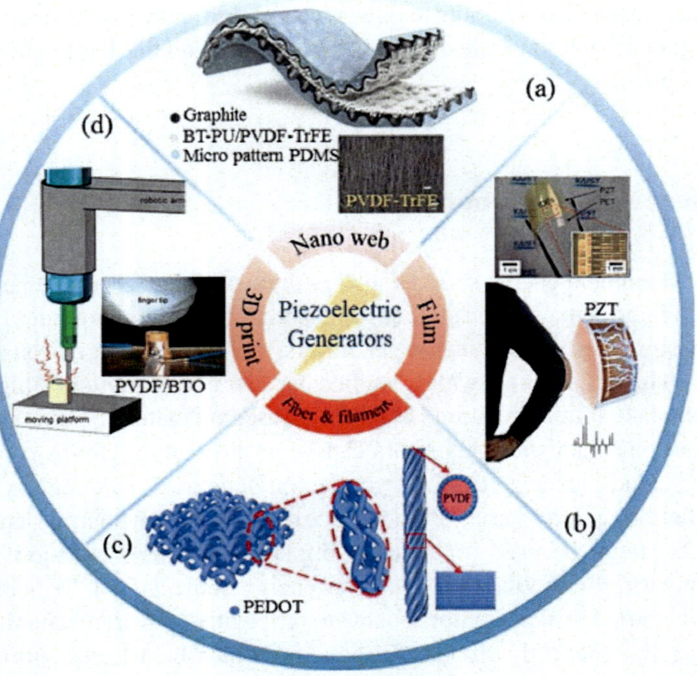

Fig. 1.3 Common fabrication methods of piezoelectric generators: **a** Schematic representation of the nanofiber stretchable piezoelectric nanogenerators with a stacked nanofiber mat and graphite electrodes [19]. **b** A schematic illustration of wearable applications using the flexible PZT energy harvester [127]. **c** Schematic illustrations for preparation of PEDOT@PVDF nanofiber woven fabric [157]. **d** Solvent evaporation-assisted 3D printing process for PVDF nanocomposite-based 3D structures [62]

1.3.1 Principle of Piezoelectricity

Piezoelectricity come from the Greek word of *piezein* which means press (squeeze) [158]. The Curie brothers was first discovered the piezoelectric effect in 1880 [158]. It is expected that more than 30% of the material available in the world have piezoelectricity effect. Despite a wide variety of materials show this property, just a few of them have found practical applications [159]. The term 'piezoelectricity' is used to explain dielectric materials that generate electrical charges across their boundaries due to the mechanical stress application, which is known as the direct piezoelectric effect (Fig. 1.4). The direct piezoelectric effect can be implemented for energy harvesting. The charge generated at each surface of materials due to the polarised aligned dipoles and domains. Ions and electrons as free charges flow through surface charge of the material [160]. In the converse (or indirect) piezoelectric effect, a mechanical strain happens when the material is subjected to an electrical field [161]. The converse piezoelectric effect is more applicable to acoustic emitter, vibration damping and actuator applications. In piezoelectric materials, the relationship between the applied stress and induced charges per unit area is linear and reversible [161]. There is an accurate definition of piezoelectricity based on crystallography and ions distribution within an single unit cell. The common case is when there is no applied pressure the centres of the negative and positive charges are non-centrosymmetric within the unit cell planes [162].

Among the 32 crystallographic point groups, 11 groups are centrosymmetric and include an inverse centre. In this case, polar properties in a material are impossible since an existing symmetry transformation inverted each polar vector. All other remaining 21 non-centrosymmetric point groups (without an inversion centre) exhibit piezoelectric properties [163]. In fact, the piezoelectric effect comes from inadequate symmetry in the ions distributions of crystalline materials [160]. Many researches have been conducted on harvesting energy from piezoelectric materials along with the closely related sub-classes of pyroelectrics and ferroelectrics [3, 164]. Figure 1.5 shows the relationships between ferroelectrics, pyroelectric, piezoelectric and dielectrics materials. Whereas these properties are available in the same material, it provides the material fascinating prospect for harvesting energy from various sources such as vibration, light and thermal variations [165].

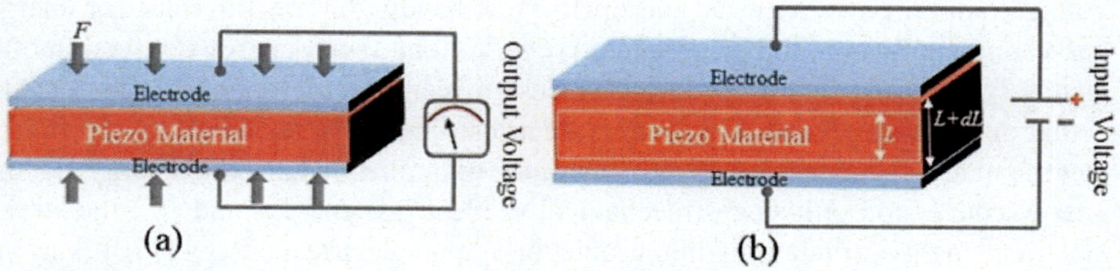

Fig. 1.4 Schematic of **a** direct and **b** converse piezoelectric effects

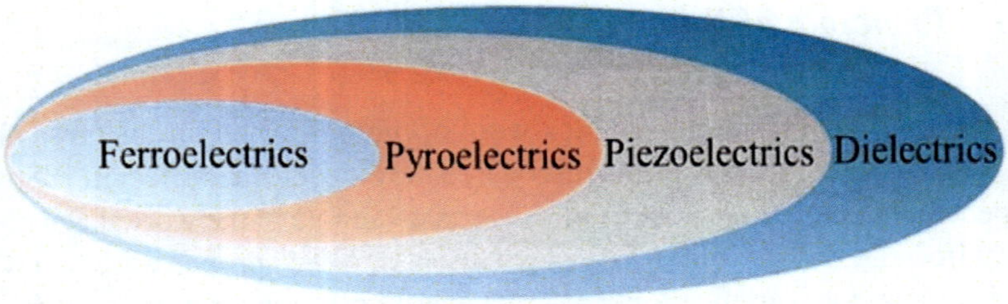

Fig. 1.5 Dielectric, piezoelectric, pyroelectric and ferroelectric material relationships

1.3.1.1 Piezoelectric Constitutive Equations

The piezoelectric constitutive equations can be given in the following for the relationship between the strain x, electric displacement D, electric field E and stress X:

$$x = s^E X + dE \tag{1.1}$$

$$D = dX + \varepsilon_0 \varepsilon^X E \tag{1.2}$$

where s^E and ε^X are the elastic compliance (under constant E) and relative permittivity (under constant X), d is the piezoelectric constant, and ε_0 is the vacuum permittivity. Note that the converse piezoelectric effect (Eq. 1.1) and the direct piezoelectric effect (Eq. 1.2) have the same piezoelectric constant [166]. The piezoelectric constitutive equations assume that the transducer strain is the sum of mechanical strain induced by the mechanical stress and the controllable actuation strain caused by the applied electric voltage [167].

1.3.1.2 Piezoelectric Figures of Merit

A variety of performance measurements have been used to compare piezoelectric materials on various applications. The piezoelectric strain constant and piezoelectric voltage constant are proper criterion for actuating and sensing application. The cost, voltage output, power density and operational bandwidth are variables for energy harvesting application to be compared [168]. Among them, energy conversion efficiency is the most important factor to design piezoelectric energy harvesters for actual applications [169]. The output electrical energy generated from the piezoelectric material depends on (i) the magnitude of applied mechanical energy, (ii) the energy conversion efficiency (mechanical to electrical energy) and (iii) the stored electrical energy productivity into usable, or output, electrical energy [170]. Various figure of merits have been generated for assessing the energy conversion efficiency of piezoelectric materials.

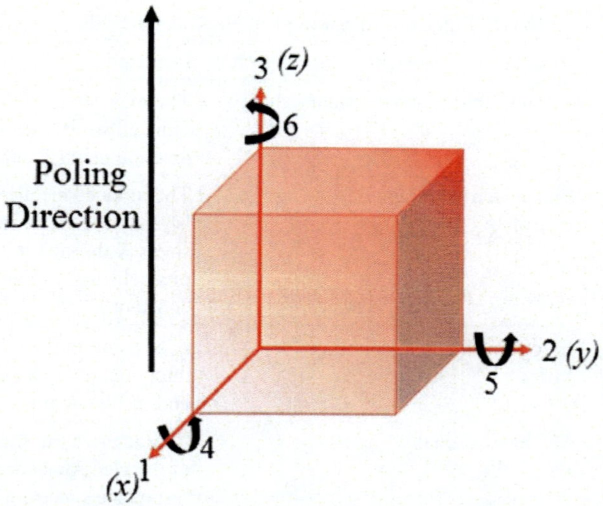

Fig. 1.6 Direction index of electromechanical constants in poled piezoelectric materials

Piezoelectrics have five significant figures of merit: (1) the piezoelectric voltage constant g, (2) the piezoelectric strain constant d, (3) the mechanical quality factor Q_m, (4) the electromechanical coupling factor k, and (5) the acoustic impedance Z. These figures of merit are presented in Table 1.1 [171]. Obviously for energy harvesting applications the figures of merit revealed that high piezoelectric activity is needed for maximum piezoelectric coefficients (such as d_{33} and d_{31}).

Figure 1.6 presents the direction indexes of piezoelectric constants in rectangular crystallographic system. The principal properties along X, Y and Z axes are described using 1, 2 and 3 notations, respectively. The shear constants are represented by 4, 5 and 6, respectively. The 33-mode refers to the application of stress in the direction of polarisation, which other modes utilising for energy harvesting [172]. In fact a superscript index on the electromechanical constant define the constant condition under conducted measurements [173].

1.3.1.3 Measuring and Enhancing Piezoelectric Coefficient (d_{33})

The d_{33} coefficient is defined as the charge produced for an applied stress, or the strain for an applied voltage, and these are theoretically equivalent. Measurement of the d_{33} coefficient can be realised in several ways varying in accuracy and simplicity [174]. A number of alternative static and quasistatic techniques had been reported to measure d_{33}.

These methods can be divided into two categories. One is to monitor the induced charge by applying a controllable external mechanical stress to the piezoelectric structure through the direct piezoelectric effect; the other is to measure the displacement in response to an applied voltage via the converse piezoelectric effort [175]. The d_{33} piezoelectric constant is approximately twice that of d_{31}, so higher performance is expected from a device operating in d_{33} mode such as pressure sensors [176].

Table 1.1 Figures of merit in piezoelectric materials

Piezoelectric constant	Symbol	Description	Equation
Piezoelectric strain constant	d	The magnitude of the induced strain x by an external electric field E	$x = dE$
Piezoelectric voltage constant	g	The induced electric field E is related to an external stress X through the piezoelectric voltage constant g	$E = gX$
Electromechanical coupling factor	k	A measure of the interchange of electrical and mechanical energy	$k^2 = \frac{\text{stored electrical energy}}{\text{input mechanical energy}}$
Mechanical quality factor	Q_m	Characterize the sharpness of the electromechanical resonance spectrum	$Q_m = \frac{\omega_0}{2\Delta\omega}$
Acoustic impedance	Z	Acoustic energy transfer between two materials	$z = \sqrt{\rho c}$

The d_{33} value can be measured by P–E loop obtained from PFM test or directly measure by d_{33} meter. The electrical energy produced by the piezoelectric generator is proportional to its piezoelectric constant [177].

PVDF and P(VDF-TrFE)s exhibit the strongest piezoelectric responses ($d_{33} \approx -30$ pC N^{-1}) among the polymer materials [178]. There is a higher d_{33} value of nanofibrous PVDF film due to the improvement in β-phase content of the PVDF polymeric material. Also the annealing treatment on nanofibrous film PVDF has exhibited high d_{33} piezoelectric coefficient [179]. The d_{33} profile along the diameter direction has the higher values at the edge of the PVDF nanofibers [180]. Controlling relative humidity and voltage polarity increased the d_{33} piezoelectric coefficient for PVDF fibers by more than three times and allowed us to generate a power density of 0.6 μW cm^{-2} from PVDF membranes [181]. The piezoelectric coefficient (d_{33}) is found to increase from 10 to 35 pC N^{-1} with an increase in the volume fraction for 70 vol% of the micron-sized $K_{0.5}Na_{0.5}NbO_3$ (KNN) crystallites in PVDF matrix [182].

1.3.2 *Piezoelectric Materials*

Piezoelectric materials have received great interest in energy harvesting technology due to their unique ability to take advantage the ambient vibrations energy conversion to electric potential [183]. The specific energy transduction of piezoelectric materials enables their applications in the area of energy harvesting, sensors, actuators, structural health monitoring, and use in health care devices [184].

Piezoelectric materials can be categories in three groups: piezoelectric polymers, piezoelectric ceramics, and piezoelectric composites [185]. Depend on the desired vibration amplitude and frequency, the piezoelectric materials can be in the form of polycrystalline ceramics, textured ceramics, polymers or thin films [186]. There are diverse piezoelectric ceramic forms, including nanowires, bulk, and nanofibers. The bulk form of piezoelectric ceramics has excellent characteristics but no flexibility, limit its use in wearable devices. On the other side, nanowires have good flexibility without mass production capability [187]. In 1950s, the discovery of the barium titanate (BT) and lead zirconate titanate (PZT) and the family of these piezoelectric materials, is viewed as a major development in the field of sensors [188].

PZT is the common ceramic based piezoelectric material. Since it contains 60% lead which is harmful for surrounding environment and human being makes the use of this material problematic. The toxicity of these ceramic material is only further enhanced due to the volatilization at high temperatures during calcination and sintering. The high densities of these materials lead to large characteristic acoustic impedances, necessitating the usage of matching layers. The large relative permittivity of these materials facilitates electrical tuning and reduce their piezoelectric voltage coefficients. The large mechanical quality factor requires the addition of damping layers to optimise resonant to an acceptable level [188]. The solution of this problem is using lead free ceramic based piezoelectric materials like BT, potassium sodium niobate ($KNaNbO_3$) and potassium niobate ($KNbO_3$) [64]. Nevertheless, piezoelectric ceramics have intrinsic limitations like lower piezoelectric voltage constant and brittleness as compared to piezoelectric polymer [189]. In the field of piezoelectric ceramics, PZT and associated materials, which exhibits piezoelectric response and high dielectric constant are most widely used. However, the application of these piezoelectric ceramics restricted owing to lead-containing, hardness, high thermal conductivity, fragility and relatively high cost. Barium Titanate as inorganic filler due to its ease in handling and environmentally friendly nature is preferable than lead-free material. However, poor flexibility of BT limits its applications for large scale fabrication. To overcome these issues, piezoelectric polymers are a good alternative to piezo ceramics. Piezoelectric polymers have the advantage of flexibility, easy fabrication process and comparatively low-priced. Therefore, fabrication of flexible sensors, fabric sensor and fiber based generators facilitate by piezoelectric polymers [190].

To improve performance, the polymer/ceramic composite has been developed as a substitute material that has the properties of both materials to provide a new material with a tailored performance as either actuating or sensing. Choosing the proper ceramic fillers is important to improve the mechanical or dielectric properties of the polymer composite [191]. Hence, polymer/ceramic composites have been investigated as candidate dielectric materials. Recently, PVDF/BT nanocomposites are highly attractive to energy harvesting and sensor applications due to their simple and accessible fabrication methods, low cost, and ideal piezoelectric properties [172].

1.3.2.1 Piezoelectric Polymer

PVDF as a most favourable piezoelectric polymer has been highlighted in material science research due to the discovery of the piezoelectric effect in PVDF films by Kawai in 1969 [190]. PVDF is a thermoplastic semi-crystalline polymer that has been extensively considered for its piezoelectric properties, moderately good chemical resistance, strength and thermal resistance [192, 193]. PVDF and its copolymers such as, poly(vinylidene fluoride-co-chlorotrifluoro ethylene, poly(vinylidene fluoride-co-trifluoroethylene), and poly(vinylidene fluoride-co-trifluoroethylene-co-chlorotrifluoro ethylene) [194] are known to have very stable and large piezoelectric, ferroelectric, and pyroelectric properties among polymers. Also, it has high resistivity against cleaning agents, skin products, and sweat [195]. The PVDF homopolymer contains 59.4 wt% fluorine, 3 wt% hydrogen. The existence of fluorine atoms with a large van der Waals radius and electronegativity in the polymer chain [$–CH_2–CF_2–$] leads to a dipole moment perpendicular to the monomer chain [160]. The piezoelectric properties of PVDF derive from the larger difference in electronegativity between carbon and fluorine atoms than between carbon and hydrogen atoms. Among a larger polymer structure, depend on the crystal geometry, the dipole moments may sum to provide a net electric dipole moment [196]. PVDF includes at least four different crystalline phases, having the non-polar α phase and polar β, δ and γ phases.

The PVDF β phase has the highest polarizability of the polar phases, but it is not appropriate for high energy storage due to ferroelectric switching and dissipating a large amount of energy during the charge-discharge process [197]. Different chain conformations leads to formation different phases: trans planar zigzag (TTT) for the β-phase, T3GT3G′ for γ and ε phases and TGTG′ (trans-gauche–trans-gauche) for the α and δ phases [198]. In particular, the β phase has intrinsic spontaneous polarization in compare with the ε and γ phases, which gives PVDF piezoelectricity and ferroelectricity [199]. Although PVDF polymer is recommended for piezoelectric applications, it has some inherent limitations including lower dielectric constant and piezoelectric charge constant (d_{33}). The piezoelectric properties of the PVDF polymer rely on the aggregate of β phase formation [200].

Enhancing Method to Increase β-phase Formation of PVDF

Over the last years, many researchers have focused on controlling the morphology and performance enhancement of PVDF relating to its piezoelectricity. Therefore, many interesting information concerning the relationship between PVDF morphology and fabrication conditions have been presented [201]. The PVDF piezoelectric properties depend upon the aggregate of β phase formation. Therefore, the β phase formation of PVDF is one of the most vital parameters for piezoelectric property, which is generally happen by subsequent stretching and high voltage poling process (Fig. 1.7) [202]. Poling process reoriented the dipole moments and leads to enhance the power generation capability. Prior to poling, for the same length of chain, the dipole moment of the stretched chain was much lower than the dipole moment of a β-PVDF chain.

When an electric field was applied to the stretched chain (over 50 MV/cm) perpendicularly, a rotation of $–CF_2$ and $–CH_2$ around the chain axis in the reverse directions is expected. When the electric field is over 100 MV/cm, the direction shift from α-phase to β-phase can be predicted [203].

The techniques that include stretching, such as electrospinning, can cause higher β phase than α phase. Due to the strong elongation flow in the jet, PVDF dipole chains align to the applied stress direction and consequently phase transformations happen by polymer chains orientation [204]. During manufacturing process of melt spun fiber, the PVDF is stretched and the polymer is mainly present in the β phase [205]. Also in the electrospinning process the β-phase content increased by optimizing the electrospinning process and decreasing the fiber diameter [206–209]. Temperature has a significant role in the PVDF crystallization procedure. Drawing at low temperature does not provide proper β phase formation and piezoelectric properties due to the random orientation [210].

The highest β phase content was achieved for the annealed PVDF film at 90 °C [211]. Thermal processing had a direct effect on modifying the crystalline microstructure and also confirmed that heat stretching and annealing could increase the degree of crystallinity and β-phase crystalline for PVDF wet spun fibers [212]. The β-phase crystallization at a high cooling rate is still controversial, but this important finding means that the crystallization mechanisms of PVDF into the β-phase could be controlled from the melt state with the cooling rate [213]. It is observed that the melt-quenching process assists the β-phase crystallization in the PVDF film (51%) and thereby enhances the domain switching behaviour, ferroelectric and dielectric properties [214]. The β-phase crystalline content in the bicomponent fibers of PVDF/PU remained unchanged after the hydrothermal treatment [215]. Current research has considered nanofillers addition such as metal nanoparticles, carbon nanotube, nanoclay, graphene, zinc oxide, barium titanate into PVDF for the β-phase formation enhancement [216]. Filler addition disrupts the polymer molecular chain, which leads to the higher piezoelectric response by enhancement of the β-phase crystallinity.

The piezoelectric response of the polymer nanocomposite increase while, ceramic based materials with high piezoelectric coefficients, like barium titanate have been incorporated into the PVDF matrix [217]. The degree of crystallinity of the PVDF/PEO films exhibits non-linear increase, with increasing amount of PEO in the films [218]. The –OH groups on cellulose as a requirement for β-phase evolution in PVDF [219]. The incorporation of functional fillers such as carbon-based materials into the PVDF matrix can induce the PVDF electroactive crystalline phase by interfacial interaction, and also enhance the dielectric constant of the composites based on the percolation theory [220].

PVDF composites with graphene oxide (GO) or reduced graphene oxide (RGO) using spin coating and annealing were found to crystallize the PVDF in ferroelectric phase. The existence of hydrogen bonding and/or electrostatic interaction between the oxygen functionalities of RGO (or GO) and PVDF causes to the nucleation of ferroelectric β or γ phase [221]. Dielectric constant increased by four times and

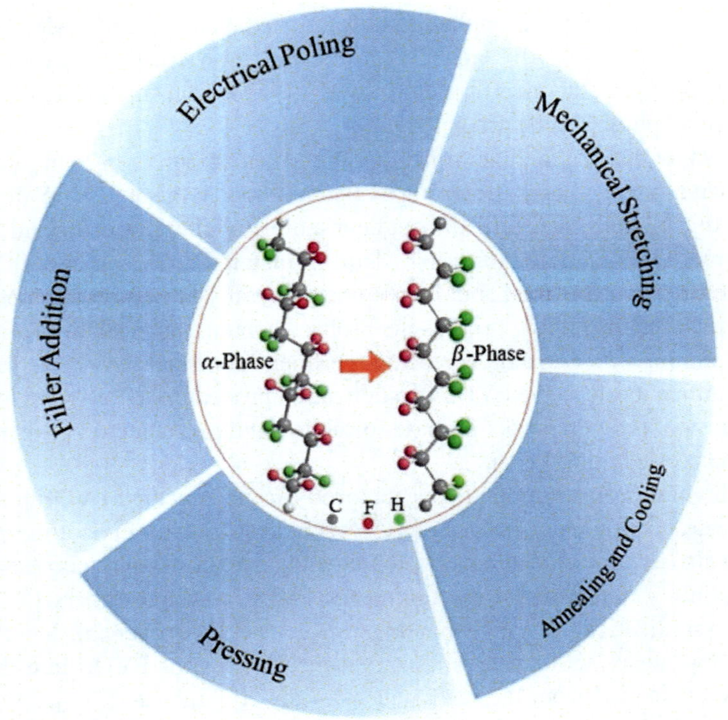

Fig. 1.7 Schematic representations of ways for β phase enhancement in PVDF-based polymers

piezoelectric coefficient also increased twice by the addition of RGO in the PVDF films.

The phase transition from α to β will contribute to an increase in overall dielectric constant upon addition of RGO [222]. Addition of GO/rGO can promote the ß-phase formation in PVDF matrix by successful interaction between –C=O group of GO/rGO and the $-CF_2$ or $-CH_2$ group of PVDF. GO and rGO possess large specific surface areas and carrier mobility coupled with abundant defects and several functional groups (−OH, –C–O–C, and –C=O groups).

The presence of these defects and abundant surface functional groups in/on GO and rGO plays a significant role in improving the dipole polarisation and dielectric properties of constituent composites [223]. In fact, the π-electrons which are mobile in RGO attracts CH_2 dipoles of PVDF and helps in aligning the β-phase of PVDF. Also addition of RGO helps in providing a conducting path for the charges induced inside the film to move to the surface and further allow the dipoles of PVDF to align which leads to enhanced piezo response [224]. In PVDF/CNT composites the enhancement and stabilization of pure β phase is possible with the CNTs proper modification [225].

The β-phase content increased by designing core-shell structure using inorganic $BaTiO_3$ nanoparticles as the core and PVDF doped with grapheme oxide nano sheets as the shell [226]. Also the β-phase content improved by doping Ag nanowires in the PVDF fibers through electrostatic interaction between surface charges of the nanowires and dipoles of the PVDF chains [227]. The pressure-quenching process could provide an ideal crystallization condition to obtain β extended chain crystals without causing undesired autonomous PVDF thermal degradation [228]. Incorporation of various nanofillers into PVDF matrix resulting variation in piezoelectric responses due to different percentages of the β-phase. However, the β-phase formation is not the only effective parameter in the piezoelectric response. The interconnectivity and the arrangement of the neighbouring dipoles are also effective in the piezoelectric response [217].

Piezoelectric Ceramic

Piezo ceramics have the best piezoelectric performance; nevertheless, their inherent brittleness restricts their application in the flexible devices. Among different ceramics, barium titanate ($BaTiO_3$), lead zirconium titanate (PZT), bismuth strontium titanate ($BiSrTiO_3$), strontium titanate ($SrTiO_3$), calcium titanate ($CaTiO_3$), PZT possesses highest dielectric and ferroelectric properties. Recently most concern is about the lead toxicity in PZT containing devices which are harmful for human health and environment [59].

To solve this problem lead free BT is considered due to the most environmentally friendly and providing high piezoelectric and ferroelectric properties along with a high dielectric constant [229].

Barium titanate with the general formula ABO_3 called perovskites, is a primitive cubic, with larger cation-A in the corner, smaller cation-B in the middle and the anion, commonly oxygen, in the centre of the face edges [230]. BT as a perovskite crystal at temperatures lower than the Curie temperature placed in a strong DC electric field the crystal undergoes a ferroelectric transition from cubic to tetragonal phase (Fig. 1.8) and becomes spontaneously polarized due to the unidirectional intermittent displacement of Ti^{4+} ions to neighbouring O^{2-} ions [231]. Barium titanate can be prepared by various methods based on the desired properties [232]. Bulk BT is fragile and therefore cannot be instantly used as mechanical energy harvesting materials [233, 234]. In order to improve the piezoelectric properties of lead-free nanomaterials, many research groups have performed single crystal growth, defect control and agglomeration prevention, which have offered reliable dielectric, piezoelectric and elastic properties, respectively [235]. Many researches have confirmed that the ceramics grain size plays an important role in its piezoelectric performance. The grain size of BT ceramics has an important effect on d_{33} [236]. It was revealed that the BT dielectric constant increased with the decrease of the average grain size firstly, reached the maximum in the particle size range of 0.8–1.1 μm, and then decreased rapidly as the grain size decreased in further [237, 238]. Nowadays, the piezoelectric performance at low mechanical impact was enhanced through the composite materials which are formed by the integration of organic polymer and inorganic material. Furthermore, the use of composite materials to fabricate piezoelectric generators devices provides several benefits including high flexibility, ease process, low internal leakage current,

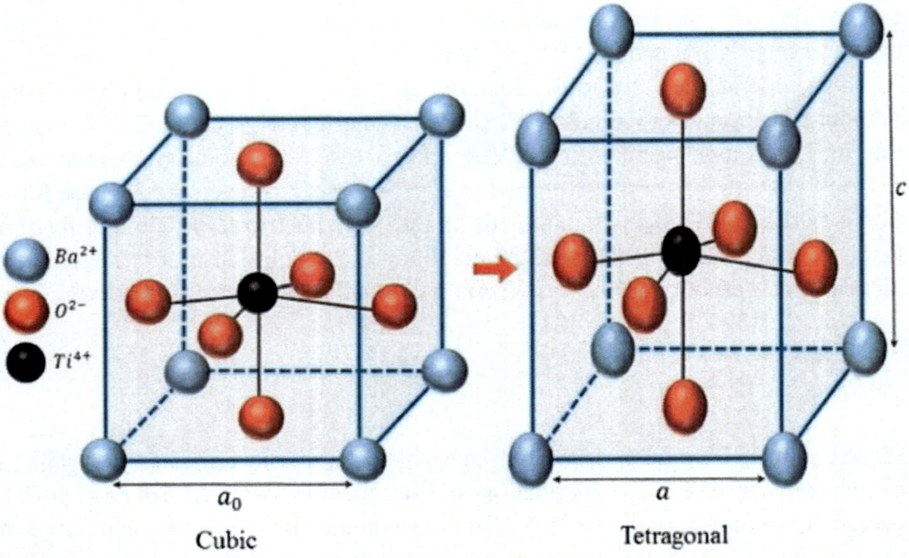

Fig. 1.8 Unit cell change of a $BaTiO_3$ crystal during spontaneous polarization

Table 1.2 Physical and piezoelectric properties of PVDF and $BaTiO_3$ at room temperature.

Materials	d_{33} (pC/V)	dielectric constant ε_r	g_{33} (10^{-3} V_m/N)	Curie temperature (°C)	Density (g/cm^3)
PVDF	30 [189]	9.2 [244]	339 [245]	80 [246]	1.78
$BaTiO_3$	190 [247]	1200 [245]	14.1 [245]	116 [248]	6.02

low cost ability to sustain/harvest large mechanical forces, mass production and higher electrical output in compare with conventional piezoelectric generators [239].

By incorporated one-dimensional BT nanowire to the nanocomposite generator film can reach both performance enhancement and homogeneous distribution of piezoelectric materials and also prevent the aggregation of piezoelectric nanoparticles [240]. The effect of BT on the electrical properties of polymer composites has attracted more attentions. It is shown that BT could increase the resistance of composites up to 186 kΩ [241]. The paraelectric BT nanoparticles modified the Young's modulus, piezoelectric coefficient, and dielectric constant of the composite [242]. The dispersion of BT nanoparticles in the PVDF matrix was greatly improved owing to the changing the BT nanoparticles by air plasma and enhance phase formation to 80% [243]. Low temperature plasma alter the BT nanoparticles surface in order to improve the interfacial compatibility between ferroelectric PVDF and BT nanoparticles. This modification demonstrated a lower dielectric loss and higher dielectric constant [243]. Table 1.2 compared physical and piezoelectric properties of PVDF and $BaTiO_3$ as piezoelectric polymer and ceramic.

1.3.3 *Piezoelectric Textile*

1.3.3.1 Charge Generation Mechanism

The applied pressure on clothes has been classified into the two groups: vertical pressure (clothing weight) and horizontal pressure (surface and contact pressure). Horizontal pressure caused by body pressure during clothing wear is crucial factor for evaluating restriction of wearer mobility. Stretchy parts in garment such as knee, elbow and back area are considered mostly for energy harvesting applications [249]. The applied strain and piezoelectric constant are two important considerations in output power generation. In comparison with bulk ceramics, PVDF has lower piezoelectric constant and can generate higher output power if external force applies induce stress in the PVDF. The finite element modelling and use of theory of curved piezoelectric generators have been used to analyse the applied stress and performance of the piezoelectric generators [250–252]. The simulation results indicate that the output voltage increase linearly as the result of bending velocity increase. The generator output current has the same trend as the output voltage. By increasing the external load, the output current decreases but the output voltage follows an opposite trend [253].

The ferroelectric domains tend to align along the direction of applied electrical field when subject to a strong electrical field, as shown in Fig. 1.9i. While there is no

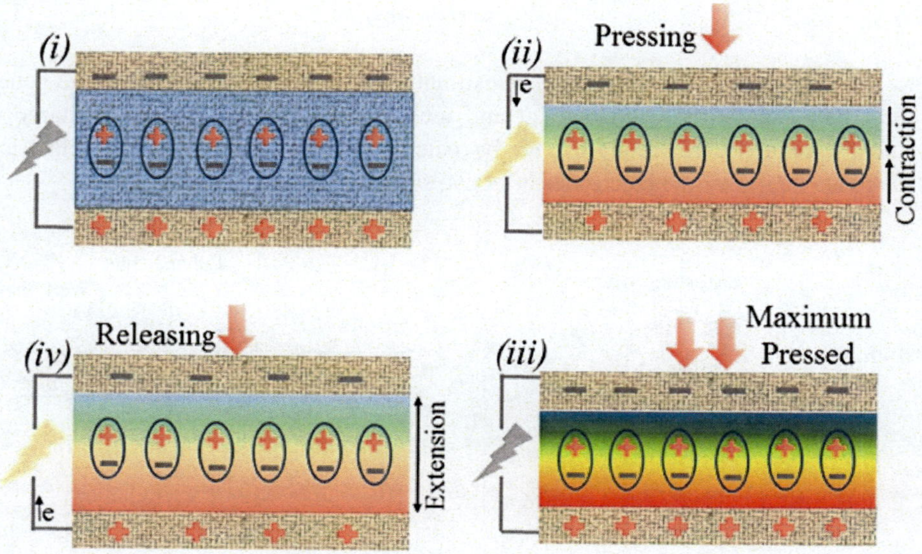

Fig. 1.9 Schematic illustration of piezoelectric charge generation process for textile base generators: **i** The dipole get align in the direction of the applied electric field after poling process. **ii** The piezoelectric potential is generated by applying compressive force. **iii** Maximized press state with highest polarization density. **iv** Electrons flow back along the opposite direction as the compressive force is released

external force on the device, the device has no output electric signal due to electrical equilibrium within the device [254]. When force is applied on the piezo fabric surface, its volume decreases and negative strain is produced. The piezo potential between the electrodes formed as a result of the charge centres separation and forming electrical dipoles. During pressing process, piezo potential force electrons to move toward external circuit and consequently reach a new equilibrium state (Fig. 1.9ii). As a result, electricity is generated from mechanical energy. The maximum pressure on the surface leads to full contact between electrodes and active area which provide highest polarization density (Fig. 1.9iii). By releasing external force, electrons flow back to the equivalence charge state in the short-circuit condition (Fig. 1.9iv) [2].

The piezoelectric coefficient (d_{3i}) as the ratio of the open circuit charge density to the applied stress is used to measure the material piezoelectric performance. Typically, the d_{33} coefficient is higher than the d_{31} coefficient [255].

While the piezoelectric energy harvesting device is subjected to compressive stress by external force, the PVDF and substrate are subjected to tensile stresses and generate electric potential (Fig. 1.10a) [253]. As an example curve theory of piezoelectric material Fig. 1.10b showed a woven textile structure with warp and weft yarns that crossing over. The weft yarns that cross the warp yarns in an orthogonal direction are produced from PVDF film and metal electrodes [256]. According to Fig. 1.10a the following Eq. (1.3) is used to calculate the output voltage (v) in mode d_{31}, where t is the thickness of the piezoelectric material, g_{31} is voltage coefficient and σ is stress.

$$V_{31} = \sigma_{XX} \cdot g_{31}.t \tag{1.3}$$

Also, the electrical charge can be estimated from the curved piezoelectric generator using electrical displacement and mechanical strain. The surface charge of the piezoelectric material, q, can be calculated from the integral of electrical displacement over the electrodes effective surface area (Fig. 1.10b) [257]:

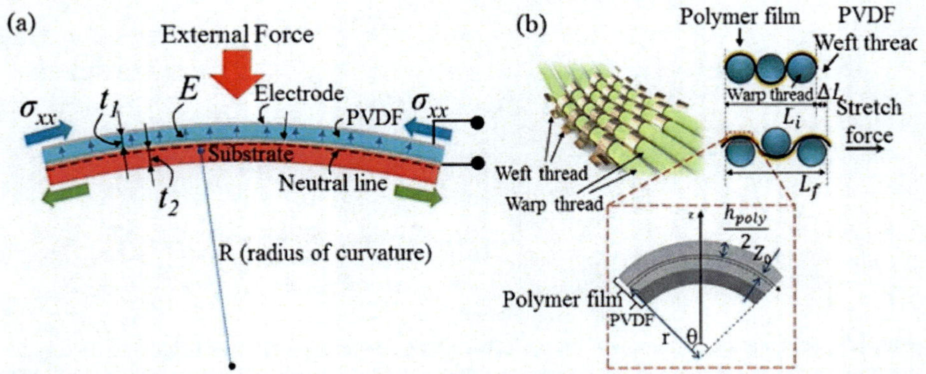

Fig. 1.10 **a** Curved piezoelectric generator with the d_{31} mode [253]. **b** The cross-sectional view of the woven structure of PVDF film and the operating principle [256]

$$q = \int_0^A D_3 . dA \tag{1.4}$$

The surface charge is defined as a function of the strain induced in the system. Thus, the electric charge depends on the strain induced in the device [253]. Based on Eq. (1.4) the electric charge per unit area affected by the strain, if the strain is increased through the garment design, an increase in the value of Q can be expected. For a harmonic excitation, the current (I) is:

$$I = \frac{dQ}{dt} \rightarrow Q = \int I dt \tag{1.5}$$

The garment design plays a significant role in increasing the efficiency of piezoelectric energy harvesters. Garment optimal design adjust the desire mechanical deformation to support performance of piezoelectric generator. In fact by increasing garment elastic activity the piezoelectric energy harvester has better performance [137].

1.3.3.2 Performance Enhancement of Piezoelectric Textiles

The various attempts to generate piezoelectric textile energy generators follow the same principles as for other piezoelectric systems and are summarised in Fig. 1.11.

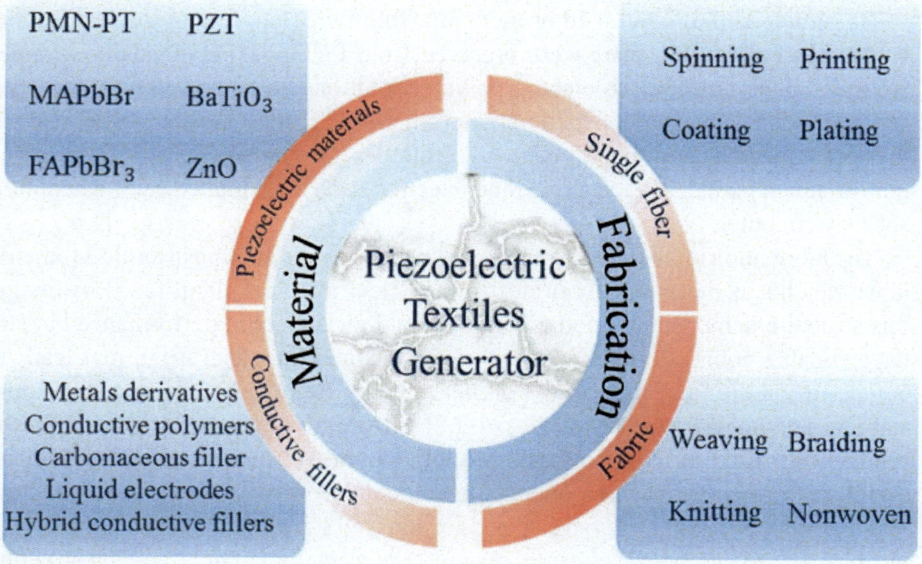

Fig. 1.11 Schematic of diverse enhancing methods for fiber based piezoelectric generators performance

Inorganic fillers in the composite structure presents lots of probable methods to enhance performance of nanocomposites, including modulating the content, size and morphology of the nanofillers. But the effect of nanofillers in piezoelectric performance and the best fabrication design of nanocomposite fiber for piezoelectric generators are still outstanding issues [258]. Selection of conductive materials with the best electrical output, ideal mechanical characteristics and long-term environmental quality is an important consideration about textile based generators [2]. The PVDF and conductive fillers (carbon black [259], carbon nanotube (CNT) [260], and graphene nanoplate [261]) combination can obtain the dielectric properties enhancement with reduce fillers consuming.

1.4 Wearable Piezoelectric Generators

1.4.1 Wearable Piezoelectric Generators Based on Piezoelectric Polymers

Fiber based piezoelectric generators with the ease of integration into the garments can directly convert mechanical energy into electrical signals. The surface charges can be transferred and collected by harvesting interface circuit. In the single fiber structure in order to collect the charge from the piezoelectric part electrodes should be in coaxial or core-shell structures. To achieve this approach conductive materials as electrodes need to be coated [262] or wrapped [263] around active piezoelectric part.

By simple rolling and coating method a four stage process piezoelectric fiber was made, electrospun mats were prepared from PVDF–TrFE manually wrapped around a silver coated piezoelectric polymer solution can be coated on conductive substrate such as metal deposited polymer fibers and metal wires but issues such as the poor adhesion between polymer and electrodes need to be resolved [267]. Nylon multifilament yarn that acted as the inner electrode and the wrapped carbon nanotube sheets were outer.

The fabrication of fiber generators were finalised by dip coating in an eleastometric layer which was mechanically protective and electrically insulator [263]. Although this structure demonstrate good mechanical and piezoelectric performance but still have slidable connection between the electrodes and the electrospun mat leads to contact instability and increased wearing, thereby negatively affecting the output stability and durability of the devices [268]. Also multiple processing steps for the production is time and cost consuming. Another improved method towards achieving a high-performance fiber in a one-step continuous method is a core shell structure which conductive material as an electrode in core and shell fabricated the same time of piezo part fabrication. Piezoelectric generator presented in a three-layered structure which consist of PVDF electrospun web sandwiched between two membranes layer of PVDF/rGO through a continuous process [265].

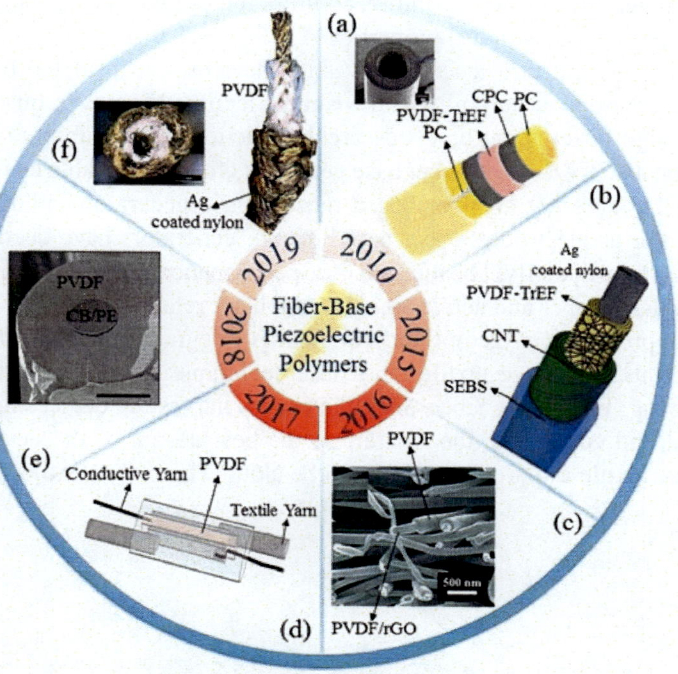

Fig. 1.12 Comparison difference types of fiber-base piezoelectric polymer: **a** A cylinder shaped of piezoelectric P(VDF-TrFE) fibre with CPC/indium as electrodes and poly(carbonate) (PC) as protective layer [264]. **b** A flexible piezoelectric fiber from electrospun PVDF-TrFE web [263]. **c** SEM image of the PVDF nanofiber membrane and the PVDF/rGO electrode [265]. **d** Conceptual illustration of design of piezofilm yarn sensor [266]. **e** SEM image of the cross-section of a single PVDF filament with conductive core electrod [4]. **f** A triaxial braid structure from melt-spun PVDF filament and conductive yarn [71]

In this technique there is a robust adhesion between the electrodes and the PVDF nanofiber that provide more polarization possibilities. Electrospinning method is not appropriate for mass production and electrospun fibers cannot be used in conventional textile industry machine to fabricate wearable textile. To solve this problem melt spinning technique with the advantage of mass production, high forming quality and on time forming has been widely used [81]. The bicomponent fibres produced with PVDF as core sheath and carbon black/polyethylene as core which are fed through two separate extruders during spinning process. Silver paste on the yarn surface consider as outer electrode [4].

Since the assembly process has a key role in performance enhancement of the piezoelectric devices, a major aim of the present study was to explore textile processing methods for making such devices. As described in more detail in Chapter three, we have produced a triaxial melt spun PVDF fiber has been developed with flexible yarn as electrodes. First, the as-spun PVDF filaments were braided around silver coated nylon yarn as inner electrodes and then the whole structure was covered with braided silver coated nylon fibers as outer electrodes. The developed fiber improved mechanical (i.e. flexibility, comfort and durability) and piezoelectric (i.e. power output and sensitivity) properties [71].

All-fiber piezoelectric energy harvesting devices, encapsulated by a protective layer to increase their durability and wearable facility. Moreover, this technique has attracted significant attention of researchers due to great feasibility of light weight design and also try to use all fiber base generators for the wearable technology [269]. Figure 1.12 represent all fiber based piezoelectric generator without using metal wires or foil as electrodes. Fiber-based energy generators have been developed in fabrics form particularly. The piezoelectric performance of the single fiber generators owing to the number and active area limitation are relatively low.

One approach to overcome this problem is integration piezoelectric fibers into fabrics by using diverse textile fabrication technique including knitting, braiding, and weaving (Fig. 1.13). Since wearable energy harvesting devices need to provide right feel and comfort to the wearer, so the best design of the device is integration in the textile as part even whole of the cloth. Therefore, textile-fiber structures

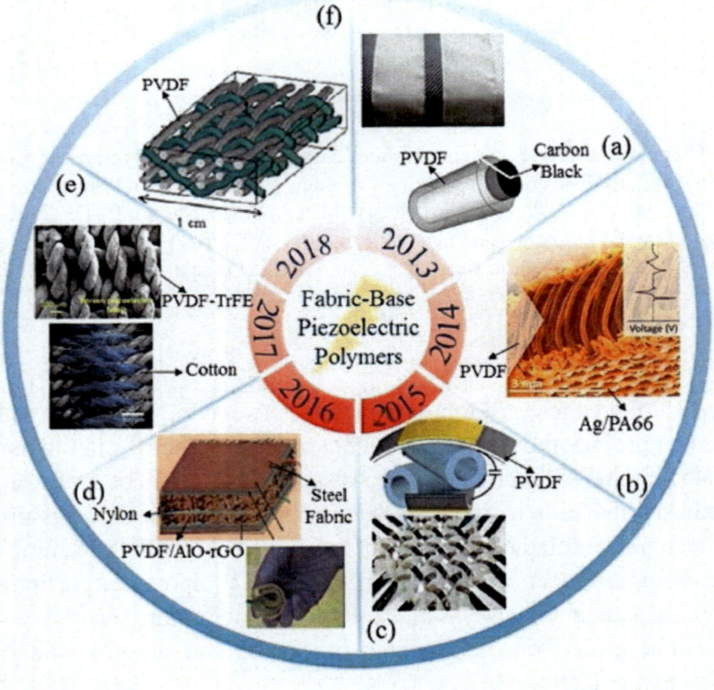

Fig. 1.13 Comparison of different types of wearable fabric-base piezoelectric polymer: **a** Fibers woven into a textile from core-shell structure of PVDF filament [271]. **b** A 3D piezoelectric fabric from PVDF filament in the knit structure [81]. **c** Woven flexible textile structure from elastic tubes and piezoelectric film bands with electrods on both sides [272]. **d** PVDF/AlO-rGO flexible nanocomposite [273]. **e** A woven piezoelectric fabric from twisted yarns of electrospun PVDF-TrFE [274]. **f** A 3D interlock woven structure established from PVDF and conductive fibers [70]

would provide perfect building elements for a smart garment, as they could be naturally integrated into fabrics during the fabrication process without affecting flexibility, comfort and air permeability [20]. The soft and flexible fiber-based generators can be designed for the high fatigue resistance under numerous deformation cycles [270]. Fabric generators should be stretchable same as bending ability to certify a fitting, improve comfortability and increase the accessibility in human motions for the recovery of energy. High strain elasticity of stretch fabrics associated with knitted fabric stretching and bending of elastic fibers, such as elastane, which offers a restoring force. Flexibility and stretchability are needed for a strong piezoelectric fiber to be incorporated into smart textiles [263]. The power output from the woven fabric of polyamide yarn and melt-spun PVDF filament showed that the harvested energy from the size 15 $\times$ 100 mm of this textile is sufficient to power low-power electronics [271].

The output voltage of 3D interlock woven fabric from 100% melt-spin piezoelectric filament was 16 times higher than the output voltage provided by a 2D plain woven [70]. The 3D spacer piezoelectric fabrics have the advantage of efficient charge collection due to proper electrodes connection and also pressure uniform distribution on the fabric surface, causing performance enhancement. Also, this all fibre piezoelectric fabric can be cut into any size and shape without affecting on its flexibility and provide a simple route for integrating [81]. Generally, 3D fabric structures in compare with 2D textiles structures have higher efficiencies because of piezoelectric fibers density [2].

Many researchers described flexible wearable generators, however many of them especially in case of electrospinning were not flexible and did not bring the right feel of comfort for the wearers. In fact, they have metal/metallic electrodes which limits their use, flexibility and lifetime [81]. The poor resistance to fatigue causes early failure of the metal foils electrodes. The separation in the metal-insulator-metal structure is due to huge mismatch between the Young's modulus and Poisson's ratio of the metal electrode and the piezoelectric layer. Therefore, loss of mechanical integrity and electric connectivity occurred during long time device operations [270]. In addition to piezoelectric materials for wearable technology, generate energy can be achieved with triboelectric materials. Triboelectricity through friction, convert mechanical energy to electricity, by a coupled effect of triboelectrification and electrostatic induction [76].

With appropriate structure designs and using effect of both piezoelectric and triboelectric effect the contact friction and the deformation of the materials can generate voltage and the applied mechanical energy is exploited with high efficiency (Fig. 1.14). The current methods to design and fabricate triboelectric structures are based on reactive ion etching or photolithography which both of them are expensive and complicated [275]. The electrospinning is a simple, versatile and low-cost method to fabricate ultrathin fibers from a wide range of materials as well as ceramics, polymers and composites which most triboelectric generators are made by this technique [68].

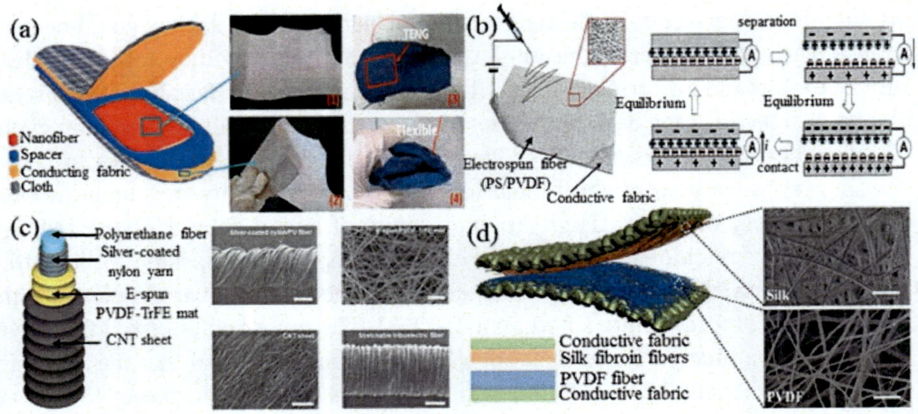

Fig. 1.14 Improving performance of fiber base generators by using piezoelectric and triboelectric materials: **a** Schematic of the flexible structure of the TENG-based insole [68]. **b** Diagram of a triboelectric energy harvester fabricated by the electrospinning process [276]. **c** Schematic of stretchable triboelectric structure and SEM image of its component [277]. **d** Schematic diagram of all-fiber hybrid-triboelectric nanogenerator with two conductive fabrics

1.4.2 *Wearable Piezoelectric Generators Based on Piezoelectric Composite*

The piezoelectric performance of fiber-based generators improved by incorporating piezoceramic nanoparticles into piezoelectric polymer matrix. PVDF and its copolymer due to their transparency, flexibility, chemical stability, biocompatibility and high breakdown strength are considered to be the most promising candidates for wearable technology application. In addition, PVDF has a high piezoelectric voltage constant (g_{33}) and a low dielectric constant ($\varepsilon_r < 10$) [278]. According to Eq. (1.6) the energy density (U_e) of dielectric material can be presented as follow where D is the electric displacement and E is the electric field:

$$U_e = \int_{D_{\max}}^{0} E dD \tag{1.6}$$

Therefore, both large electric displacement and high breakdown strength are desired for a high energy density. Common dielectric materials are usually modified with high dielectric constants ceramics, but in some cases undesirable temperature dependent breakdown strengths and loss characteristics could be happened [279]. To avoid these limitations, significant efforts have been made in making polymer–ceramic composites to achieve both a high breakdown strength and superior dielectric properties [280]. The amount of ceramic filler into the polymer matrix is very critical and generally the composite reach the peak dielectric properties at >50 vol% ceramic filler [280].

The addition of high-permittivity nanofillers is the most common way to enhance the dielectric properties of polymer composites. However nanoparticles incorporation into the polymer matrix can improve the polymer composites permittivity, but, it can also cause to the decrease of dielectric breakdown strength. In some composite structure's cavitation occur around the filler due to absence of adhesion between the fillers and matrix. Surface modification of the ceramic nanofillers is used to modify the interfacial areas between ceramic and polymer matrix and consequently avoid the filler aggregation and increase their dispersion in polymer matrix [281].

The usage of an appropriate surface modifier plays a key role for improving the piezoelectric response [282]. It is desirable to fabricate generators from perovskite materials with large piezoelectric coefficients including BT (100 pC/N) and Pb(Zr, Ti)O_3 (PZT, 200 pC/N) for energy harvesting [283]. Fabric-based piezoelectric generators are fabricated from traditional textile manufacturing, fabric platform method and multilayer stacking technique for electrospun webs (Fig. 1.15).

In a stack form a PVDF–$NaNbO_3$ electrospun web and elastic conductive fabric as electrodes converts compression energy of 0.2 MPa into output voltage of

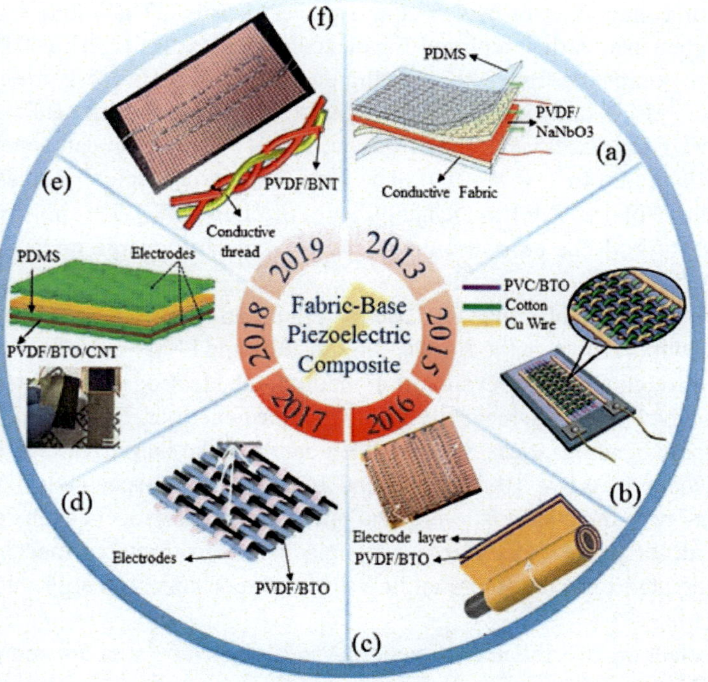

Fig. 1.15 Comparison of the different types of wearable piezoelectric generators base on piezoelectric composite: **a** All-fiber wearable nanogenerator from electrospun web of PVDF–$NaNbO_3$ fibers [270]. **b** A 2D piezoelectric fabric consisting piezoelectric fabric integrated with conducting wire electrodes and insulating spacer cotton yarn [20]. **c** Woven cotton-based textile containing four piezoelectric fibers [154]. **d** A cotton-based textile containing piezoelectric fibers in a woven structure [153]. **e** Textile-based hybrid nanogenerator from electrospun PVDF/CNT/BT web [284]. **f** Core-shell piezoelectric nanofiber yarns with external electrodes [28]

3.4 V, and short-circuit current of 4.4 μA at low frequency [270]. The nonwoven PVDF/CNT/BT fabric fabricated by electrospinning technique without further polarization. This layer embedded between two layer of conductive fabrics as electrodes to generates a rectified average peak output voltage of ~162 V [284]. The corresponding voltage and current for PVDF/BT (20 wt%) and PZT/PVDF (20 wt% PZT) fiber generators were ~6 V and ~4 nA and ~3 V and ~1.2 nA, respectively [285]. In the tri-layer structure achieves by stacking two layers of PVDF/BT nanocomposite films, one on each side of the surface-modified *n*-type graphene, a peak output voltage of 10 V along with 2.5 μA current at an applied force of 2 N generated [278]. The double-layered BT/PVDF films in addition to better flexibility and good interfacial adhesion, the mechanical property and cyclic durability can also be improved. The double-layer BT/PVDF (20 wt%) film illustrate significant performance of 6.7 V and 2.4 μA in output voltage and output current respectively [258].

The electrospun web of 15 wt% BT nanoparticles into the P(VDF-TrFE) solution can generate an output voltage of 25 V during walking at frequency of 0.6 Hz under 600 N weight [286]. Research confirmed that existence of 20% BT filler of 10 nm size in the P(VDF-TrFE) matrix harvest ~0.28 μW energy [287]. The PVDF/BT meltspun fibers are knitted to form a wearable energy generator that produced a maximum voltage output of 4 V and a power density 87 μW cm^{-3} which is 45 times higher than earlier reported for piezoelectric textiles [288]. BT/PVDF/silver nanowires composite film with avoiding electrical poling process presented a high voltage of ~14 V and current of ~0.96 μA [239]. By dispersing BT nanoparticles in the PVDF matrix, this particular film composite orientation can generate highest output voltage up to 150 V [289]. The piezoelectric coefficient increased by 47% after addition of 5 wt% BT in the nanofiber of P(VDF–TrFE). This improvement also observed for both the current outputs and piezoelectric charge under compression force [290].

On the fabric surface, piezoelectric materials can be grown and covered directly. ZnO nanowires grown on the Au/Kapton substrate and PVDF coated on the structure. Finally, through a thermal-evaporation method, the electrode was coated on the top surface of the PVDF to achieve the hybrid flexible generator. The hybrid-fiber device, under elbow bending-releasing for ~90° generated the output voltage, current and power density of 0.1 V, 10 nA cm^{-2} and 16 μW cm^{-3}, respectively [291]. Hybrid generators with the combination of two power generation mechanisms got research interest recently. Piezoelectric and triboelectric nanostructures mechanisms have been integrated onto the fibers and convert the mechanical energy into electricity (Fig. 1.16).

The selection of triboelectric material and its surface area are more critical to obtain high power output. In order to enhance the triboelectric performance, the ZnO nanoflowers as the piezoelectric material used to improve the PDMS surface roughness. The device generated electrical output voltage/current values of ~75 V/3.2 μA, ~150 V/8 μA, during walking (Fig. 1.16a) [292].

The piezoelectric nanogenerator in the hybrid structure acts as supportive for triboelectric nanogenerator for energy generation while it is not working in some situations. As a self-powered strain sensor, ZnO nanorods were grown around the

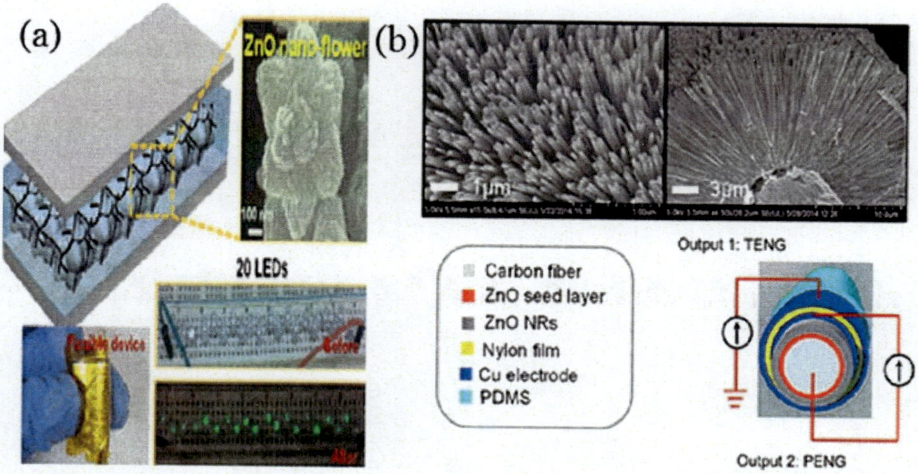

Fig. 1.16 Fiber-based hybric triboelectric and piezoelectric generators: **a** PDMS-encapsulated nanoflower-like ZnO composite film [292]. **b** Fully integrated 3D fabrication method for fiber-based hybrid nanogenerator [293]

conductive carbon fibers by hydrothermal growth method to harvest energy from human movement (Fig. 1.16b) [293]. A comprehensive overview about performance of PVDF base piezoelectric generators is presented in Table 1.3.

1.4.3 Smart Wearable Market

Wearable technology is a new trend that merges electronics to the daily activities and makes an effort to change the lifestyles. The market is divided based on product type, such as smart clothing, ear worn, smartwatch, fitness trackers, body worn, head-mounted display camera and medical devices. The aim of wearable technologies is to improve physical performance and form regular physical activity and develop new habits through digital persuasion. Wearable electronic devices may offer social influencing or gamification of the exercise through challenges creation as well as virtual rewards to enhance physical performance [319]. Since, wearable technology industry is a relatively new concept, the landscape is crowded with vendors across many market areas in the world. The market has already advanced beyond human's fitness and wellness, health care and commercial and industrial applications. The market is expected to grow to an enormous value of 160 billion USD by 2026 from its current value of 40 billion USD [141]. Amongst the 423 wearable device products on the current market, the largest market share associated with fitness and lifestyle (Fig. 1.17a) [320]. The world market for wearable devices, now consisting of wireless headphones with smart assistants, increased 15.3% over the last year to 198.5 million

Table 1.3 Extensive comparison between piezoelectric performances of fiber based PVDF generators.

Material	Structure	Signals	Excitation	Active area	Wearable	References
PVDF	Single fiber	30 mV, 3 nA	Strain at 2 Hz	–	✖	[294]
PVDF	Fiber	76 mV, 39 nA, 577.6 pW cm^{-2}	Strain of 0.05% at 7 Hz	–	✖	[295]
PVDF/BTZO	Hybrid composite film	11.9 V, 1.35 μA, 15.8 nW	11 N cyclic pushing	2.5 cm × 2.5 cm	✖	[296]
PVDF/GO	Nanofiber	2 V, 29 mW cm^{-2}	Bending, 2.5 N	7 × 5 × 0.009 cm^3	✖	[297]
PVDF	Film	1.25 V, 1.79 μW cm^{-2}	1.2 N at 3 Hz	45 × 45 mm	✖	[298]
PVDF/ZnO	Film	1.81 V, 0.57 μA, 0.21 μW/cm^2	15 kPa	1.5 × 1.5 cm	✖	[61]
P(VDF–TrFE) /BaTiO3	Film	0.28 μW	Bending	–	✖	[287]
PVDF	Nanofiber	3.50 V	Bending	25 × 40 mm^2	✖	[299]
PVDF-TrFe	Nanofiber	1.5 V, 40 nA	Bending	–	✖	[300]
P(VDF-TrFE)	Fiber	4.0 V, 2.6 μA	Impact 25 N	10 mm × 10 mm	✖	[301]
PVDF/nanoclay	Nanofiber	70 V, 68 μW cm^{-2}	Tapping	–	✖	[302]
PVDF/$BaTiO_3$	Nanofiber	331 nA	Squatting up and down	2 cm × 1 cm	✖	[303]
PVDF/KNN	Filament	3.7 V, 0.326 μA	Finger Tapping	–	✖	[64]
PVDF	Filament	2.3 V	2.2 lb from a height of 5.5 cm	–	✖	[304]
PVDF	Filament	2.2 V	1.02 kg form a height of 5 cm	–	✖	[305]
PVDF	Film	22 V, 28.5 μW/cm^2	Finger tapping	0.7 × 0.7 cm^2	✖	[306]

(continued)

Table 1.3 (continued)

Material	Structure	Signals	Excitation	Active area	Wearable	References
PVDF	Film	560 V, 3.7 W/m^2	Hand clapping	7 cm × 5 cm	✖	[307]
PVDF-TrFE/BNT-ST	Nanofiber	0.99 V	Vibration	–	✖	[308]
PVDF	Film	11 V	2.95 N	2 mm × 2 mm	✖	[309]
MgO/P(VDF-TrFE)	Film	2 V	Bending	7 × 7 mm^2	✖	[310]
P(VDF-TrFE)	Film	200 nA, 30 V	0.588 MPa	5 cm × 2 cm × 0.3 cm	✖	[311]
PVDF	Nanofiber	48 V	8.3 kPa	2 × 3 cm^2	✖	[312]
$BaTi_2O_5$/PVDF	Film	27.4 μW/cm^3, 35 V	Vibration	–	✖	[313]
PVDF	Nanofiber	7 V, 700 nA, 0.85 mW/m^2	5 N	5 cm^2	✓	[314]
PVDF	Fiber	Sensor	Deflection	190 mm × 70 mm × 2.8 mm	✓	[205]
PVDF	Filament	700 mV	Impact of motor	–	✓	[315]
PVDF/ZnO	Fiber	0.1 V, 10 nA cm^{-2}, 16 μW cm^{-3}	Bending arm	–	✓	[291]
PVDF-TrFE	Nanofiber coil	20 mV	Bending	306 μm	✓	[72]
PVDF-TrFE	Yarn (Braid)	500 mV, 0.412 mVm/N	1 MPa at 0.5 Hz	–	✓	[74]
PVDF + GO + AlO	Nanocomposite film	36 V, 0.8 μA, 27.97 μW cm^{-3}	Human finger tapping	–	✓	[273]
PVDF	Fabric	112.7 V, 80 mW/m^2	Bending	16 cm^2	✓	[316]
PVDF	3D fabric	5.10 μW cm^{-2}, 14 V, 30 μA	Pressure 0.10 MPa	5 cm × 6 cm	✓	[81]

(continued)

Table 1.3 (continued)

Material	Structure	Signals	Excitation	Active area	Wearable	References
P(VDF-TrFE)/$BaTiO_3$	Hybrid fiber	8.8 μW cm^{-2}, 10 V, 3.5 μA	0.02 MPa	–	✓	[317]
PVDF	Filament	8 V, 4 μW	Shoulder strap of a laptop case	2.5 cm × 20 cm	✓	[4]
PVDF-HFP-TEA·BF	Yarn	3.5 V, 43 μWh cm^{-2}	Bending	–	✓	[318]
PVDF/CB	Filament	4 V, 15 nW	0.07% strain, 0.05 N	25 mm	✓	[271]
PVDF	Filament	380 mV, 29.62 μW cm^{-3}	0.023 MPa impact	0.6 mm × 2 cm	✓	[71]
PVDF-TrFE	Fiber	24 mV, 8 nA	50% strain	440 μm × 10 mm	✓	[277]
PVDF	Filament	2.3 V, 10.5 μJ m^{-2}	5 N	–	✓	[70]
(PVDF-TrFE)/BNT-ST	Yarn	1.9 V, 0.38 μA, 0.88 μW	Body pressure	256 cm	✓	[28]
PVDF	Nanofiber	210 V, 2.1 mW, 45 μA	Body pressure	5 cm × 6 cm	✓	[68]
PVDF/$NaNbO_3$	Nanofiber	3.4 V, 4.4 μA	0.2 MPa	2.5 × 2.5 × 0.2 cm	✓	[270]
PVDF/CNT/$BaTiO_3$	Nanofiber	161.66 V, 2.22 W m^{-2}	580 N	4.5 × 5 cm	✓	[284]

parts in the late of 2019. Driving that growth will be the continued spread of ear-worn devices, watches and wrist bands as well as further adoption in the healthcare division.

The market is expected to achieve 279 million units by the end of 2023 with an 8.9% annual growth rate. Since 2018–2022 the wristband type growth will remain muted with a rate of 0.3%. However, it is important to mention that this category will still represent for 24.7% of the total market by 2022 with the total amount reaching 47.0 million [320]. These wearable devices will continue to be multi-purpose from stepping stone performing into the wearables market for first time buyers to serving as simpler alternatives for smartwatches. In advance markets, the low cost of wearable devices make them possible to serve as simple patient health monitoring equipments [321]. The overall wearables market had a 17% annual growth in shipping a total of 897,000 units in the third quarter of the year 2018 from the India. Suppliers will be looking to investigate some of the new categories for the smart watches with cellular connectivity to provide the solutions around healthcare and lifestyle management [322]. Over the year 2019, 85 million smartwatches has been sold which will increase to 137 million in 2022 but that fitness trackers market dropped by 43 million to 39 million in comparison to the last four years. Predictions suggest that the global smart fashion market will reach 2.9 billion USD till 2022, based on the wide range of fashion clothes design (tunic, dresses, etc.) and accessories (jewellery, watches, shoes, bags, etc.) [97]. Huawei posted the largest annual growth among the market-leading vendors (Xiaomi, Fitbit, Garmin, Apple) presenting a renewed fitness bands with several new models and an introduction to smart earwear with its sport pulse earphones that track heart rate as well as audio [323]. Currently, Samsung, Google, Apple, HUAWEI and other companies have associated smart devices. Their products have their own features including: health monitoring, information reminder, map navigation, connecting smart home and independent call [324]. Smart textiles market is foreseen to develop due to the integration of traditional clothing industry and electronics to create textiles and connected apparels for application in drug delivery, fitness, and clinical trials (Fig. 1.17b) [325].

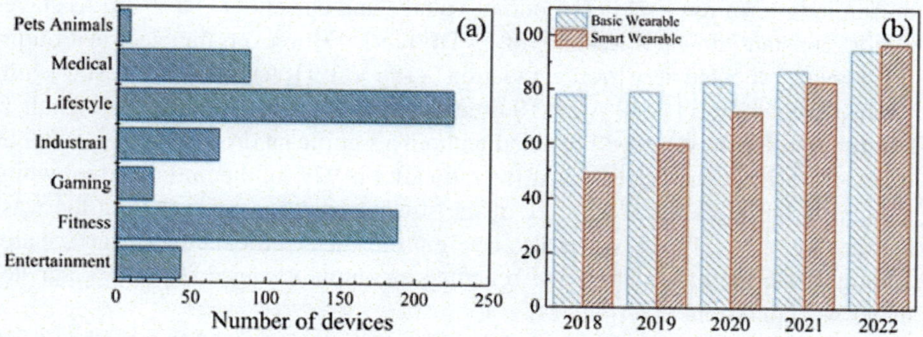

Fig. 1.17 **a** Wearable device products on the market. **b** Worldwide wearable forecast by product category

1.4.4 Washable Wearable Generators

Electronic systems from conventional semiconductor materials, are not ideal choice for textiles integration due to incompatible properties, including rigidity, bulk and related problems to wearability and lack of washing durability [326]. Therefore, the all fiber based generators are designed for wearable electronic devices usage. The significant factors including wicking properties, stretchability, air permeability and recovery require testing and control to provide user comfort. A general question concerned about wearable electronics and all types of e-textiles is their washability [327]. Textiles like any other materials when are used for outdoor applications will be exposed to the natural elements and required cleaning/washing at some point. Accordingly, the washing process effects need to be investigated in detail for the polymeric piezoelectric wearable devices [109]. Washability is one more key concern for the electronic textile in real personal application which expected to be similar to commonly worn clothes and for all fiber base generator this ability is highly demanded. The influence of wear and tear, washing cycles and daily use also should be monitored to ensure piezoelectric response accuracy and provide a certain lifetime value for the fabric [81]. The piezoelectric behaviour of polyamide-11, polypropylene and PVDF concerning capability of peak-to-peak voltage generation after washing at 40 °C with the addition of detergent was investigated. The results showed that after washing process the PA11 and PVDF yarn present higher peak-to-peak output voltage in compare with the pristine specimen (Fig. 1.18a) [315]. A simple CF_4 plasma etching followed by water immersion of the electrospun PVDF mats caused self-cleaning properties and super hydrophobicity. This would increase the durability of the smart textile and reduce the washing cycles when it is used. The established electrospun PVDF mat revealed ~31% improvement in piezoelectric properties after surface modification (Fig. 1.18b) [299]. Melt-spun piezoelectric PVDF fibers with conductive cores were attached onto a textile glove to operate as a sensor feature. The washing effect on the electrical resistance of the printed interconnections showed that following 15 wash cycles the average surface resistivity was still below 500 Ω/square, which was supposed to be adequate for the operation of polymeric sensor system during long-term use [328]. The output voltage and current (~110 V, 2 μA) stayed almost constant after 2 h washing of PVDF fabric. Moreover, the electrical output performance showed very low degradation (~106 V, 1.9 μA) following 12 h washing process (Fig. 1.18c) [316]. After 10 times of washing, the PVDF output signals of stitch-based sensor were well retained without a significant drop in current. After 50 times of washing, the current retention ratio after is 97% of the initial current which signifying the mechanical stability of the stitch sensor against repeating the wash cycles (Fig. 1.18d) [109]. Water also can improve the electrical contact surface area among PVDF melt spun fibres and generates an output voltage of up to 8 V per step under wet conditions [4].

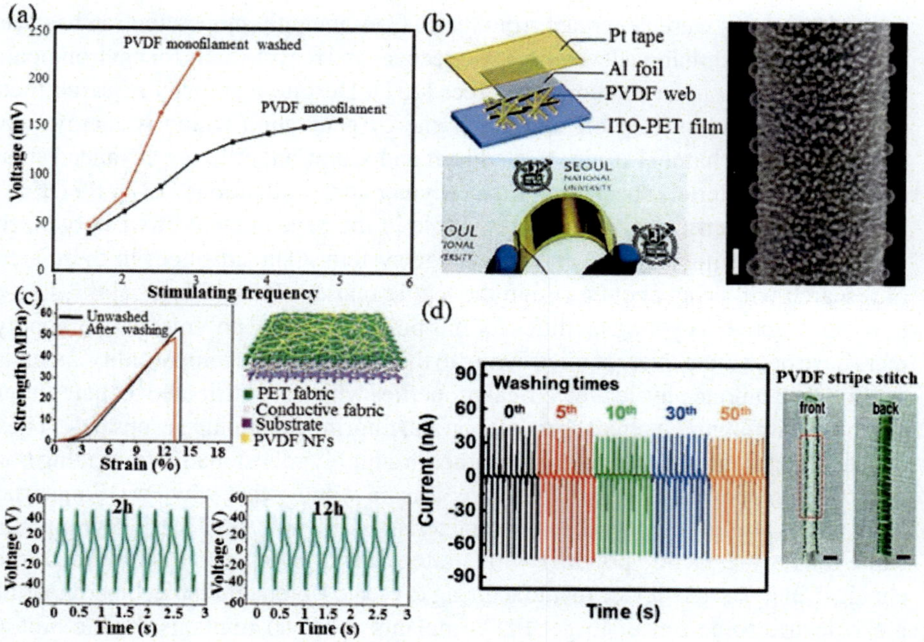

Fig. 1.18 Washability of wearable generators: **a** Peak-to-peak voltage generation of PVDF monofilaments before and after washing [315]. **b** Schematic of the nanogenerator structure and SEM image of PVDF electrospun web after plasma etching [299]. **c** A comparison of voltage output of the power generation fabrics after 2 h, 12 h washing and mechanical properties [316]. **d** Output current of the PVDF stitch-based textile sensor after 50 cycles of washing [109]

1.4.5 Degradation and Recycling of Wearable Generators

World population has been continuously and cause increasing affluence and advent of the fast fashion. This phenomenon leads to textile industry fast growth from design to delivery and makes less durable clothing which bring a dramatic increase in the textiles production consequently. Global per capita textile fiber production had ~120% growth trends over the years 1975–2018. This development in textile production is related to environmental challenges such as natural resources demands, textile production and use emissions, and the solid waste generation. The annual textile waste per person is about 40 kg and 19 kg for the United States and China, respectively [329]. Beside the fashion revolution, multitasking application of textiles such as energy harvesting and signal detecting increase demand for more textile production which highlighted the importance of textile recycling challenge. Waste electronics recycling in order to restore raw materials is beneficial for the environment, especially for electronic devices such as lithium-ion batteries (LiBs) including valuable metals (cobalt and lithium) as the electro-active materials. Recycling of these resources is an important factor for the environment and many research work in this case [330]. LiBs as an energy source contain toxic heavy metals in a high percentage. In 2005, the 4000 t of used LiBs collected which 28% and more that 5% heavy metals and toxic electrolytes were generated from them. Consequently more attention has been needed for the technology development to recover and recycle LiBs for environmental protection also conserve natural resources [331]. Despite that many research focus on approaches for sustainable and green energy generation which is energy scavenging from mechanical strain, wasted heat and vibration, utilizing advanced smart piezoelectric materials specially PVDF polymer attract attentions [297]. PVDF as a flexible piezoelectric polymer has a key role in the development of wearable technology, electric mobility and energy storage systems. The advances in these areas of research will bring about a rise in the textile market. This will reveal the urge to develop effective recycling facilities for this polymer. PVDF polymer and its copolymers have been intensively studied owing to their excellent biocompatibility, amazing stability and unique physical/electrical properties which are attributed to their unique composition comprising hydrogen, carbon and fluorine functional groups [61]. These constituent groups combine with each other in a highly robust matrix, providing both hydrogen and covalent bonds. Therefore, it is no wonder that why PVDF materials have not exhibited biodegradable characteristics till now [332]. Biodegradability came for Poly-L-lactide polymer with strategy for material processing, electromechanical analysis and device manufacturing to create a biodegradable, biocompatible piezoelectric force sensor to get FDA-implants approval implants for the biological forces monitoring such as the pressure of diaphragmatic contraction [333]. The biodegradation properties will possibly be identified for PVDF in near future. The recycled PVDF hollow fiber membranes were fabricated using dry-wet spinning technique. The decrease of PVDF molecular weight due to increasing running time also led recycled PVDF hollow fiber membranes to have higher porosity and better permeability [334]. In fiber based wearable generators beside PVDF other part such as electrodes and connectors threads are textile based and need to recycle as well. Textile recycling mostly involves the reprocessing of the textile waste for use in new textile or non-textile products. Textile recycling methods are typically categorized as being either chemical, mechanical or more rarely, thermal. Regards to the above-mentioned challenges, there is relevant interest in increasing textile recycling and reuse, which would move the treatment of textile waste further up in the waste classification, in accordance with the European Union legislation on waste streams (European Commission (EC), 2008). Textile recycling and reusing could potentially reduce the production of pure textile fibres and avoid engineering processes further downstream in the textile product life cycle, and consequently reduce environmental effects [335].

1.5 Thesis Aims

Piezoelectric fibers not only exhibited excellent flexibility, significantly enhanced piezoelectricity and durability, but also show their potential applications for scavenging biomechanical energy and in situ monitoring human movement as self-powered wearable sensors in our daily life. To date, electrospun fibers and casted film have dominated this research field, with only few studies on meltspun piezoelectric fiber emerging in recent times. Here, PVDF based meltspun fibers offer exciting properties in the wearable devices, yet their piezoelectric properties and such properties in general in the microfiber scale have not been studied. Understanding these properties is critical to fabricating structures, posttreatment process, integration to the textile, excitation methods and harvesting energy in wearable devices with controllable and desirable properties. To address this, the thesis aims to:

Aim 1: Investigating the Properties of Piezoelectric Fiber at the Microscale
Although piezoelectric polymer (PVDF) have been extensively studied in nanofiber and film structure, fewer studies have focused on their piezoelectric behaviour at the microscale level and their characterizations. PVDF base meltspun fibers are precursor to the energy harvesting from human motion due to their flexibility, mass production and low cost. therefore aim (1) attempts to address this by employing extruder machine to fabricate composite fiber from a range of different polymer, including PVDF, PVDF/BT, PVDF/rGO and PVDF/rGO/BT and their characterization to provide important basis for the selection of suitable piezoelectric material.

Aim 2: Investigating the Effect of Textile Structure on Piezoelectric Response of the Fibers
Having investigated integration of piezoelectric fibers with each other and with the electrodes to collect the charge is required to have generator and sensor with high output voltage and sensitivity. Aim (2) attempts to address this by employing coiling, weaving, knitting, and braiding techniques to make different textile structures. Also, different fiber-based electrodes (fiber and fabric) along with piezoelectric fibers involved in the structures to provide flexible structure for different applications which did not study before.

Aim 3: Investigating the Piezoelectric Response of Wearable Devices with Different Excitation Methods
As the development wearable device continues to emerge, it will be important to understand how to excite their structures to get better response. Piezoelectric structures depend on the used material (polymer or ceramic) are sensitive to the excitation methods to present their maximum ability. Aim (3) attempts to address this by using different excitation methods including impact test, stretching, and bending in various strains and frequencies. The wearable device embedded in the garment structure and evaluated during human activity (walking and running). Also, the durability of the responses assessed after cycles of loading.

References

1. Shirvanimoghaddam, M., Shirvanimoghaddam, K., Abolhasani, M.M., Farhangi, M., Barsari, V.Z., Liu, H., Dohler, M., Naebe, M.: Towards a green and self-powered internet of things using piezoelectric energy harvesting. IEEE Access **7**, 94533–94556 (2019)

2. Dong, K., Peng, X., Wang, Z.L.: Fiber/fabric-based piezoelectric and triboelectric nanogenerators for flexible/stretchable and wearable electronics and artificial intelligence. Adv. Mater. **0**(0), 1902549
3. Narita, F., Fox, M.: A review on piezoelectric, magnetostrictive, and magnetoelectric materials and device technologies for energy harvesting applications. Adv. Eng. Mater. **20**(5), 1700743 (2018)
4. Lund, A., Rundqvist, K., Nilsson, E., Yu, L., Hagström, B., Müller, C.: Energy harvesting textiles for a rainy day: woven piezoelectrics based on melt-spun PVDF microfibres with a conducting core. NPJ Flex. Electron. **2**(1), 9 (2018)
5. Tao, X.: Study of fiber-based wearable energy systems. Acc. Chem. Res. **52**(2), 307–315 (2019)
6. Bouzelata, Y., Kurt, E., Uzun, Y., Chenni, R.: Mitigation of high harmonicity and design of a battery charger for a new piezoelectric wind energy harvester. Sens. Actuators A **273**, 72–83 (2018)
7. Gareh, S., Kok, B., Yee, M., Borhana, A.A., Alswed, S.: Optimization of the compression-based piezoelectric traffic model (CPTM) for road energy harvesting application. Int. J. Renew. Energy Res. (IJRER) **9**(3), 1272–1282 (2019)
8. Chatterjee, P.P.: Integration of solar and wind energy to generate piezoelectric potential difference. J. Resour. Energy Dev. **16**(1), 1–8 (2019)
9. Elsheikh, A.H., Sharshir, S.W., Elaziz, M.A., Kabeel, A., Guilan, W., Haiou, Z.: Modeling of solar energy systems using artificial neural network: a comprehensive review. Sol. Energy **180**, 622–639 (2019)
10. Hanif, A., Diao, S., Pei, H., Li, Z., Sun, G.: Green lightweight laminated cementitious composite (LCC) for wind energy harvesting—a novel application of LCCs. Case Stud. Constr. Mater. **10**, e00217 (2019)
11. Cansiz, M., Altinel, D., Kurt, G.K.: Efficiency in RF energy harvesting systems: a comprehensive review. Energy 174, 292–309 (2019)
12. Yuan, M., Cao, Z., Luo, J., Zhang, J., Chang, C.: An efficient low-frequency acoustic energy harvester. Sens. Actuators A **264**, 84–89 (2017)
13. Sultana, A., Alam, M.M., Middya, T.R., Mandal, D.: A pyroelectric generator as a self-powered temperature sensor for sustainable thermal energy harvesting from waste heat and human body heat. Appl. Energy **221**, 299–307 (2018)
14. Elmoughni, H.M., Menon, A.K., Wolfe, R.M.W., Yee, S.K.: A textile-integrated polymer thermoelectric generator for body heat harvesting. Adv. Mater. Technol. **4**(7), 1800708 (2019)
15. Wang, W., Xu, J., Zheng, H., Chen, F., Jenkins, K., Wu, Y., Wang, H., Zhang, W., Yang, R.: A spring-assisted hybrid triboelectric–electromagnetic nanogenerator for harvesting low-frequency vibration energy and creating a self-powered security system. Nanoscale **10**(30), 14747–14754 (2018)
16. Ryu, H., Yoon, H.-J., Kim, S.-W.: Hybrid energy harvesters: toward sustainable energy harvesting. Adv. Mater. **31**(34), 1802898 (2019)
17. Chatterjee, P., Bryant, M.: Aeroelastic-photovoltaic ribbons for integrated wind and solar energy harvesting. Smart Mater. Struct. **27**(8), 08LT01 (2018)
18. Shi, B., Liu, Z., Zheng, Q., Meng, J., Ouyang, H., Zou, Y., Jiang, D., Qu, X., Yu, M., Zhao, L., Fan, Y., Wang, Z.L., Li, Z.: Body-integrated self-powered system for wearable and implantable applications. ACS Nano **13**(5), 6017–6024 (2019)
19. Siddiqui, S., Lee, H.B., Kim, D.-I., Duy, L.T., Hanif, A., Lee, N.-E.: An omnidirectionally stretchable piezoelectric nanogenerator based on hybrid nanofibers and carbon electrodes for multimodal straining and human kinematics energy harvesting. Adv. Energy Mater. **8**(2), 1701520 (2018)
20. Zhang, M., Gao, T., Wang, J., Liao, J., Qiu, Y., Yang, Q., Xue, H., Shi, Z., Zhao, Y., Xiong, Z., Chen, L.: A hybrid fibers based wearable fabric piezoelectric nanogenerator for energy harvesting application. Nano Energy **13**, 298–305 (2015)
21. Huang, L., Lin, S., Xu, Z., Zhou, H., Duan, J., Hu, B., Zhou, J.: Fiber-based energy conversion devices for human-body energy harvesting. Adv. Mater. **0**(0), 1902034

22. Fernández-Caramés, T.M., Fraga-Lamas, P.: Towards the Internet of smart clothing: a review on IoT wearables and garments for creating intelligent connected e-textiles. Electronics **7**(12), 405 (2018)
23. He, T., Shi, Q., Wang, H., Wen, F., Chen, T., Ouyang, J., Lee, C.: Beyond energy harvesting—multi-functional triboelectric nanosensors on a textile. Nano Energy **57**, 338–352 (2019)
24. Kim, J., Campbell, A.S., de Ávila, B.E.-F., Wang, J.: Wearable biosensors for healthcare monitoring. Nat. Biotechnol. **37**, 1 (2019)
25. Torah, R., Lawrie-Ashton, J., Li, Y., Arumugam, S., Sodano, H.A., Beeby, S.: Energy-harvesting materials for smart fabrics and textiles. MRS Bull. **43**(3), 214–219 (2018)
26. Memarian, F., Rahmani, S., Yousefzadeh, M., Latifi, M.: Wearable technologies in sportswear. In: Materials in Sports Equipment, pp. 123–160. Elsevier (2019)
27. Hazarika, A., Deka, B.K., Jeong, C., Park, Y.-B., Park, H.W.: Biomechanical energy-harvesting wearable textile-based personal thermal management device containing epitaxially grown aligned Ag-tipped-$Ni_xCo_{1-x}Se$ nanowires/reduced graphene oxide. Adv. Funct. Mater. **29**(31), 1903144 (2019)
28. Ji, S.H., Cho, Y.-S., Yun, J.S.: Wearable core-shell piezoelectric nanofiber yarns for body movement energy harvesting. Nanomaterials **9**(4), 555 (2019)
29. Liu, J., Gu, L., Cui, N., Bai, S., Liu, S., Xu, Q., Qin, Y., Yang, R., Zhou, F.: Core-shell fiber-based 2D woven triboelectric nanogenerator for effective motion energy harvesting. Nanoscale Res. Lett. **14**(1), 311 (2019)
30. Kim, K.N., Chun, J., Kim, J.W., Lee, K.Y., Park, J.-U., Kim, S.-W., Wang, Z.L., Baik, J.M.: Highly stretchable 2D fabrics for wearable triboelectric nanogenerator under harsh environments. ACS Nano **9**(6), 6394–6400 (2015)
31. Dong, K., Deng, J., Zi, Y., Wang, Y.C., Xu, C., Zou, H., Ding, W., Dai, Y., Gu, B., Sun, B.: 3D orthogonal woven triboelectric nanogenerator for effective biomechanical energy harvesting and as self-powered active motion sensors. Adv. Mater. **29**(38), 1702648 (2017)
32. Wu, Q., Hu, J.: A novel design for a wearable thermoelectric generator based on 3D fabric structure. Smart Mater. Struct. **26**(4), 045037 (2017)
33. Proto, A., Vlach, K., Conforto, S., Kasik, V., Bibbo, D., Vala, D., Bernabucci, I., Penhaker, M., Schmid, M.: Using PVDF films as flexible piezoelectric generators for biomechanical energy harvesting. Lék. Tech. Clin. Technol. **47**(1), 5–10 (2017)
34. Mokhtari, F., Shamshirsaz, M., Latifi, M., Asadi, S.: Comparative evaluation of piezoelectric response of electrospun PVDF (polyvinilydine fluoride) nanofiber with various additives for energy scavenging application. J. Text. Inst. **108**(6), 906–914 (2017)
35. Sabry, R.S., Hussein, A.D.: PVDF: ZnO/$BaTiO_3$ as high out-put piezoelectric nanogenerator. Polym. Test. **79**, 106001 (2019)
36. Chen, T., Qiu, L., Yang, Z., Cai, Z., Ren, J., Li, H., Lin, H., Sun, X., Peng, H.: An integrated "energy wire" for both photoelectric conversion and energy storage. Angew. Chem. Int. Ed. **51**(48), 11977–11980 (2012)
37. Pu, X., Hu, W., Wang, Z.L.: Toward wearable self-charging power systems: the integration of energy-harvesting and storage devices. Small **14**(1), 1702817 (2018)
38. Han, S.A., Lee, J., Lin, J., Kim, S.-W., Kim, J.H.: Piezo/triboelectric nanogenerators based on 2-dimensional layered structure materials. Nano Energy **57**, 680–691 (2019)
39. Lee, H., Roh, J.-S.: Wearable electromagnetic energy-harvesting textiles based on human walking. Text. Res. J. **89**(13), 2532–2541 (2019)
40. He, J., Qian, S., Niu, X., Zhang, N., Qian, J., Hou, X., Mu, J., Geng, W., Chou, X.: Piezoelectric-enhanced triboelectric nanogenerator fabric for biomechanical energy harvesting. Nano Energy **64**, 103933 (2019)
41. Zhang, L., Su, C., Cheng, L., Cui, N., Gu, L., Qin, Y., Yang, R., Zhou, F.: Enhancing the performance of textile triboelectric nanogenerators with oblique microrod arrays for wearable energy harvesting. ACS Appl. Mater. Interfaces **11**(30), 26824–26829 (2019)
42. Song, G.J., Cho, J.Y., Kim, K.-B., Ahn, J.H., Song, Y., Hwang, W., Hong, S.D., Sung, T.H.: Development of a pavement block piezoelectric energy harvester for self-powered walkway applications. Appl. Energy **256**, 113916 (2019)

43. Lefeuvre, E., Audigier, D., Richard, C., Guyomar, D.: Buck-boost converter for sensorless power optimization of piezoelectric energy harvester. IEEE Trans. Power Electron. **22**(5), 2018–2025 (2007)
44. He, S., Dong, W., Guo, Y., Guan, L., Xiao, H., Liu, H.: Piezoelectric thin film on glass fiber fabric with structural hierarchy: an approach to high-performance, superflexible, cost-effective, and large-scale nanogenerators. Nano Energy **59**, 745–753 (2019)
45. Gao, T., Liao, J., Wang, J., Qiu, Y., Yang, Q., Zhang, M., Zhao, Y., Qin, L., Xue, H., Xiong, Z.: Highly oriented $BaTiO_3$ film self-assembled using an interfacial strategy and its application as a flexible piezoelectric generator for wind energy harvesting. J. Mater. Chem. A **3**(18), 9965–9971 (2015)
46. Kumari, P., Rai, R., Sharma, S., Shandilya, M., Tiwari, A.: State-of-the-art of lead free ferroelectrics: a critical review. Adv. Mater. Lett **6**(6), 453–484 (2015)
47. Yan, J., Jeong, Y.G.: High performance flexible piezoelectric nanogenerators based on $BaTiO_3$ nanofibers in different alignment modes. ACS Appl. Mater. Interfaces **8**(24), 15700–15709 (2016)
48. Dahiya, A.S., Morini, F., Boubenia, S., Nadaud, K., Alquier, D., Poulin-Vittrant, G.: Organic/inorganic hybrid stretchable piezoelectric nanogenerators for self-powered wearable electronics. Adv. Mater. Technol. **3**(2), 1700249 (2018)
49. Hossain, S.M., Uddin, M.N.: Energy harvesting from human foot movement. Int. J. Ambient Energy 1–6 (2018)
50. Jing, Q., Kar-Narayan, S.: Nanostructured polymer-based piezoelectric and triboelectric materials and devices for energy harvesting applications. J. Phys. D Appl. Phys. **51**(30), 303001 (2018)
51. Yan, J., Liu, M., Jeong, Y.G., Kang, W., Li, L., Zhao, Y., Deng, N., Cheng, B., Yang, G.: Performance enhancements in poly(vinylidene fluoride)-based piezoelectric nanogenerators for efficient energy harvesting. Nano Energy **56**, 662–692 (2019)
52. Anand, A., Bhatnagar, M.C.: Role of vertically aligned and randomly placed zinc oxide (ZnO) nanorods in PVDF matrix: used for energy harvesting. Mater. Today Energy **13**, 293–301 (2019)
53. Karumuthil, S.C., Rajeev, S.P., Varghese, S.: Poly(vinylidene fluoride-trifluoroethylene)-ZnO nanoparticle composites on a flexible poly(dimethylsiloxane) substrate for energy harvesting. ACS Appl. Nano Mater. **2**(7), 4350–4357 (2019)
54. Nguyen, D.-N., Moon, W.: Fabrication and characterization of a flexible PVDF fiber-based polymer composite for high-performance energy harvesting devices. J. Sens. Sci. Technol. **28**(4), 205–215 (2019)
55. Alam, M.M., Sultana, A., Mandal, D.: Biomechanical and acoustic energy harvesting from TiO_2 nanoparticle modulated PVDF nanofiber made high performance nanogenerator. ACS Appl. Energy Mater. **1**(7), 3103–3112 (2018)
56. Karan, S.K., Das, A.K., Bera, R., Paria, S., Maitra, A., Shrivastava, N.K., Khatua, B.B.: Effect of γ-PVDF on enhanced thermal conductivity and dielectric property of Fe-rGO incorporated PVDF based flexible nanocomposite film for efficient thermal management and energy storage applications. RSC Adv. **6**(44), 37773–37783 (2016)
57. Sarkar, S., Garain, S., Mandal, D., Chattopadhyay, K.: Electro-active phase formation in PVDF–$BiVO_4$ flexible nanocomposite films for high energy density storage application. RSC Adv. **4**(89), 48220–48227 (2014)
58. De Freitas, R.L.B., Sakamoto, W.K., Freitas, L.P.S., Castro, F., Lima Filho, A.P., Kitano, C., De Carvalho, A.A.: Characterization of PZT/PVDF composite film as functional material. IEEE Sens. J. **18**(12), 5067–5072 (2018)
59. Tabhane, G.H., Giripunje, S.M., Kondawar, S.B.: Intensifying energy density, dielectric and mechanical properties of electroactive β-PVDF/f-BTO nanocomposites. Physica B **571**, 149–161 (2019)
60. Anand, A., Bhatnagar, M.C.: Effect of sodium niobate ($NaNbO_3$) nanorods on β-phase enhancement in polyvinylidene fluoride (PVDF) polymer. Mater. Res. Express (2018)

61. Singh, H.H., Singh, S., Khare, N.: Enhanced β-phase in PVDF polymer nanocomposite and its application for nanogenerator. Polym. Adv. Technol. **29**(1), 143–150 (2018)
62. Bodkhe, S., Turcot, G., Gosselin, F.P., Therriault, D.: One-step solvent evaporation-assisted 3D printing of piezoelectric PVDF nanocomposite structures. ACS Appl. Mater. Interfaces **9**(24), 20833–20842 (2017)
63. Kim, K.M., Park, N.-G., Ryu, K.S., Chang, S.H.: Characteristics of PVdF-HFP/TiO_2 composite membrane electrolytes prepared by phase inversion and conventional casting methods. Electrochim. Acta **51**(26), 5636–5644 (2006)
64. Bairagi, S., Ali, S.W.: A unique piezoelectric nanogenerator composed of melt-spun PVDF/KNN nanorod-based nanocomposite fibre. Eur. Polymer J. **116**, 554–561 (2019)
65. Mi, H.-Y., Jing, X., Zheng, Q., Fang, L., Huang, H.-X., Turng, L.-S., Gong, S.: High-performance flexible triboelectric nanogenerator based on porous aerogels and electrospun nanofibers for energy harvesting and sensitive self-powered sensing. Nano Energy **48**, 327–336 (2018)
66. Harstad, S., D'Souza, N., Soin, N., El-Gendy, A.A., Gupta, S., Pecharsky, V.K., Shah, T., Siores, E., Hadimani, R.L.: Enhancement of β-phase in PVDF films embedded with ferromagnetic Gd_5Si_4 nanoparticles for piezoelectric energy harvesting. AIP Adv. **7**(5), 056411 (2017)
67. Ponnamma, D., Aljarod, O., Parangusan, H., Al-Maadeed, M.A.A.: Electrospun nanofibers of PVDF-HFP composites containing magnetic nickel ferrite for energy harvesting application. Mater. Chem. Phys. 122257 (2019)
68. Huang, T., Wang, C., Yu, H., Wang, H., Zhang, Q., Zhu, M.: Human walking-driven wearable all-fiber triboelectric nanogenerator containing electrospun polyvinylidene fluoride piezoelectric nanofibers. Nano Energy **14**, 226–235 (2015)
69. Hofmann, P., Walch, A., Dinkelmann, A., Selvarayan, S.K., Gresser, G.T.: Woven piezoelectric sensors as part of the textile reinforcement of fiber reinforced plastics. Compos. A Appl. Sci. Manuf. **116**, 79–86 (2019)
70. Talbourdet, A., Rault, F., Lemort, G., Cochrane, C., Devaux, E., Campagne, C.: 3D interlock design 100% PVDF piezoelectric to improve energy harvesting. Smart Mater. Struct. **27**(7), 075010 (2018)
71. Mokhtari, F., Foroughi, J., Zheng, T., Cheng, Z., Spinks, G.M.: Triaxial braided piezo fiber energy harvesters for self-powered wearable technologies. J. Mater. Chem. A **7**(14), 8245–8257 (2019)
72. Baniasadi, M., Huang, J., Xu, Z., Moreno, S., Yang, X., Chang, J., Quevedo-Lopez, M.A., Naraghi, M., Minary-Jolandan, M.: High-performance coils and yarns of polymeric piezoelectric nanofibers. ACS Appl. Mater. Interfaces **7**(9), 5358–5366 (2015)
73. Gowthaman, S., Chidambaram, G.S., Rao, D.B.G., Subramya, H.V., Chandrasekhar, U.: A review on energy harvesting using 3D printed fabrics for wearable electronics. J. Inst. Eng. (India) Ser. C **99**(4), 435–447 (2018)
74. Park, S., Kwon, Y., Sung, M., Lee, B.-S., Bae, J., Yu, W.-R.: Poling-free spinning process of manufacturing piezoelectric yarns for textile applications. Mater. Des. **179**, 107889 (2019)
75. Chai, Z., Zhang, N., Sun, P., Huang, Y., Zhao, C., Fan, H.J., Fan, X., Mai, W.: Tailorable and wearable textile devices for solar energy harvesting and simultaneous storage. ACS Nano **10**(10), 9201–9207 (2016)
76. Guo, Y., Zhang, X.-S., Wang, Y., Gong, W., Zhang, Q., Wang, H., Brugger, J.: All-fiber hybrid piezoelectric-enhanced triboelectric nanogenerator for wearable gesture monitoring. Nano Energy **48**, 152–160 (2018)
77. Yu, A., Pu, X., Wen, R., Liu, M., Zhou, T., Zhang, K., Zhang, Y., Zhai, J., Hu, W., Wang, Z.L.: Core–shell-yarn-based triboelectric nanogenerator textiles as power cloths. ACS Nano **11**(12), 12764–12771 (2017)
78. Dong, K., Wang, Y.-C., Deng, J., Dai, Y., Zhang, S.L., Zou, H., Gu, B., Sun, B., Wang, Z.L.: A highly stretchable and washable all-yarn-based self-charging knitting power textile composed of fiber triboelectric nanogenerators and supercapacitors. ACS Nano **11**(9), 9490–9499 (2017)

79. Choi, A.Y., Lee, C.J., Park, J., Kim, D., Kim, Y.T.: Corrugated textile based triboelectric generator for wearable energy harvesting. Sci. Rep. **7**, 45583 (2017)
80. Matsouka, D., Vassiliadis, S., Bayramol, D.V.: Piezoelectric textile fibres for wearable energy harvesting systems. Mater. Res. Express **5**(6), 065508 (2018)
81. Soin, N., Shah, T.H., Anand, S.C., Geng, J., Pornwannachai, W., Mandal, P., Reid, D., Sharma, S., Hadimani, R.L., Bayramol, D.V.: Novel "3-D spacer" all fibre piezoelectric textiles for energy harvesting applications. Energy Environ. Sci. **7**(5), 1670–1679 (2014)
82. Heo, J.S., Eom, J., Kim, Y.-H., Park, S.K.: Recent progress of textile-based wearable electronics: a comprehensive review of materials, devices, and applications. Small **14**(3), 1703034 (2018)
83. Bahadir, S.K., Sahin, U.K.: A wearable heating system with a controllable e-textile-based thermal panel. Wearable Technol. 175 (2018)
84. Yun, Y.J., Hong, W.G., Kim, D.Y., Kim, H.J., Jun, Y., Lee, H.-K.: E-textile gas sensors composed of molybdenum disulfide and reduced graphene oxide for high response and reliability. Sens. Actuators B Chem. **248**, 829–835 (2017)
85. Gonçalves, C., Ferreira da Silva, A., Gomes, J., Simoes, R.: Wearable E-textile technologies: a review on sensors, actuators and control elements. Inventions **3**(1), 14 (2018)
86. Ghahremani Honarvar, M., Latifi, M.: Overview of wearable electronics and smart textiles. J. Text. Inst. **108**(4), 631–652 (2017)
87. Wu, W., Haick, H.: Materials and wearable devices for autonomous monitoring of physiological markers. Adv. Mater. **30**(41), 1705024 (2018)
88. Schneegass, S., Amft, O.: Smart Textiles. Springer (2017)
89. Aroganam, G., Manivannan, N., Harrison, D.: Review on wearable technology sensors used in consumer sport applications. Sensors **19**(9), 1983 (2019)
90. Kubley, A., Chauhan, D., Kanakaraj, S.N., Shanov, V., Xu, C., Chen, R., Ng, V., Bell, G., Verma, P., Hou, X., Chitranshi, M., Pujari, A., Schulz, M.J.: Smart textiles and wearable technology innovation with carbon nanotube technology. In: Schulz, M.J., Shanov, V., Yin, Z., Cahay, M. (eds) Nanotube Superfiber Materials, 2nd edn., pp. 263–311. William Andrew Publishing (2019) (Chapter 12)
91. Trung, T.Q., Lee, N.-E.: Flexible and stretchable physical sensor integrated platforms for wearable human-activity monitoringand personal healthcare. Adv. Mater. **28**(22), 4338–4372 (2016)
92. Dehghani, M., Dangelico, R.M.: Smart wearable technologies: current status and market orientation through a patent analysis. In: 2017 IEEE International Conference on Industrial Technology (ICIT), pp. 1570–1575, 22–25 Mar 2017 (2017)
93. Hughes-Riley, T., Dias, T., Cork, C.: A historical review of the development of electronic textiles. Fibers **6**(2), 34 (2018)
94. Sharapov, V.: General information about piezoelectric sensors. In: Piezoceramic Sensors, pp. 1–24. Springer (2011)
95. Stoppa, M., Chiolerio, A.: Wearable electronics and smart textiles: a critical review. Sensors **14**(7), 11957–11992 (2014)
96. Syduzzaman, M., Patwary, S.U., Farhana, K., Ahmed, S.: Smart textiles and nano-technology: a general overview. J. Text. Sci. Eng **5**, 1000181 (2015)
97. Dominique, P., Crego, P.: Wearables, Smart Textiles and Smart Apparel. Elsevier (2018)
98. Park, S., Chung, K., Jayaraman, S.: Wearables: fundamentals, advancements, and a roadmap for the future. In: Sazonov, E., Neuman, M.R. (eds.) Wearable Sensors, pp. 1–23. Academic Press, Oxford (2014) (Chapter 1.1)
99. Persson, N.-K., Martinez, J.G., Zhong, Y., Maziz, A., Jager, E.W.H.: Actuating textiles: next generation of smart textiles. Adv. Mater. Technol. **3**(10), 1700397 (2018)
100. Koncar, V.: Smart Textiles for in Situ Monitoring of Composites. Woodhead Publishing (2018)
101. Kongahage, D., Foroughi, J.: Actuator materials: review on recent advances and future outlook for smart textiles. Fibers **7**(3), 21 (2019)
102. Tebyetekerwa, M., Marriam, I., Xu, Z., Yang, S., Zhang, H., Zabihi, F., Jose, R., Peng, S., Zhu, M., Ramakrishna, S.: Critical insight: challenges and requirements of fibre electrodes for wearable electrochemical energy storage. Energy Environ. Sci. **12**(7), 2148–2160 (2019)

103. Schneegass, S., Amft, O.: Introduction to smart textiles. In: Schneegass, S., Amft, O. (eds.) Smart Textiles: Fundamentals, Design, and Interaction, pp. 1–15. Springer International Publishing, Cham (2017)
104. Paul, R.: High Performance Technical Textiles. Wiley (2019)
105. Koncar, V.: Smart Textiles and Their Applications. Woodhead Publishing (2016)
106. Wang, L., Fu, X., He, J., Shi, X., Chen, T., Chen, P., Wang, B., Peng, H.: Application challenges in fiber and textile electronics. Adv. Mater. **0**(0), 1901971
107. Yilmaz, N.D.: Smart Textiles: Wearable Nanotechnology. Wiley (2018)
108. Wright, R., Keith, L.: Wearable technology: if the tech fits, wear it. J. Electron. Resourc. Med. Libr. **11**(4), 204–216 (2014)
109. Shin, Y.-E., Lee, J.-E., Park, Y., Hwang, S.-H., Chae, H.G., Ko, H.: Sewing machine stitching of polyvinylidene fluoride fibers: programmable textile patterns for wearable triboelectric sensors. J. Mater. Chem. A **6**(45), 22879–22888 (2018)
110. Ramasamy, S., Balan, A.: Wearable sensors for ECG measurement: a review. Sens. Rev. **38**(4), 412–419 (2018)
111. Castano, L.M., Flatau, A.B.: Smart fabric sensors and e-textile technologies: a review. Smart Mater. Struct. **23**(5), 053001 (2014)
112. Memarian, F., Rahmani, S., Yousefzadeh, M., Latifi, M.: Wearable technologies in sportswear. In: Subic, A. (ed.) Materials in Sports Equipment, 2nd edn., pp. 123–160. Woodhead Publishing (2019) (Chapter 4)
113. Shi, J., Liu, S., Zhang, L., Yang, B., Shu, L., Yang, Y., Ren, M., Wang, Y., Chen, J., Chen, W., Chai, Y., Tao, X.: Smart textile-integrated microelectronic systems for wearable applications. Adv. Mater. **0**(0), 1901958
114. Li, Z., Zhu, M., Qiu, Q., Yu, J., Ding, B.: Multilayered fiber-based triboelectric nanogenerator with high performance for biomechanical energy harvesting. Nano Energy **53**, 726–733 (2018)
115. Michael, A., Tehrani, K., Wearable Technology and Wearable Devices: Everything You Need to Know (2014). http://www.wearabledevices.com/what-is-a-wearable-device/. Accessed 27 Aug 2017 [WebCite Cache ID 6t1sQoa7o]
116. Teferra, M.N., Kourbelis, C., Newman, P., Ramos, J.S., Hobbs, D., Clark, R.A., Reynolds, K.J.: Electronic textile electrocardiogram monitoring in cardiac patients: a scoping review protocol. JBI Database Syst. Rev. Implement. Rep. **17**(2), 147–156 (2019)
117. Tsao, L., Li, L., Ma, L.: Human work and status evaluation based on wearable sensors in human factors and ergonomics: a review. IEEE Trans. Human Mach. Syst. **49**(1), 72–84 (2019)
118. Kwak, Y.H., Kim, J., Kim, K.: Sleep monitoring sensor using flexible metal strain gauge. Jpn. J. Appl. Phys. **57**(5S), 05GD03 (2018)
119. Halson, S.L.: Sleep Monitoring in Athletes: Motivation, Methods, Miscalculations and Why it Matters. Sports Medicine (2019)
120. Pagola, J., Juega, J., Francisco-Pascual, J., Moya, A., Sanchis, M., Bustamante, A., Penalba, A., Usero, M., Cortijo, E., Arenillas, J.F., Calleja, A.I., Sandin-Fuentes, M., Rubio, J., Mancha, F., Escudero-Martinez, I., Moniche, F., de Torres, R., Pérez-Sánchez, S., González-Matos, C.E., Vega, Á., Pedrote, A.A., Arana-Rueda, E., Montaner, J., Molina, C.A., Pagola, J., Juega, J., Francisco-Pascual, J., Moya, A., Sanchis, M., Bustamante, A., Penalba, A., Usero, M., Cortijo, E., Arenillas, J.F., Calleja, A.I., Sandin-Fuentes, M., Rubio, J., Mancha, F., Escudero-Martinez, I., Moniche, F., de Torres, R., Eichau, S., González-Matos, C.E., Vega, Á., Pedrote, A.A., Arana-Rueda, E., Montaner, J., Molina, C.A., Muchada, M., Rodriguez-Luna, D., Rodriguez, N., Sanjuan, E., Rubiera, M., Boned, S., Ribó, M., Montiel, E., Beato-Coelho, J., González Alujas, T., Evangelista, A.: Yield of atrial fibrillation detection with textile wearable holter from the acute phase of stroke: pilot study of crypto-AF registry. Int. J. Cardiol. **251**, 45–50 (2018)
121. Li, J., Ma, Q., Chan, A.H.S., Man, S.S.: Health monitoring through wearable technologies for older adults: smart wearables acceptance model. Appl. Ergon. **75**, 162–169 (2019)
122. Jin, H., Huynh, T.-P., Haick, H.: Self-healable sensors based nanoparticles for detecting physiological markers via skin and breath: toward disease prevention via wearable devices. Nano Lett. **16**(7), 4194–4202 (2016)

123. Awolusi, I., Marks, E., Hallowell, M.: Wearable technology for personalized construction safety monitoring and trending: review of applicable devices. Autom. Constr. **85**, 96–106 (2018)
124. Gilshteyn, E.P., Amanbaev, D., Silibin, M.V., Sysa, A., Kondrashov, V.A., Anisimov, A.S., Kallio, T., Nasibulin, A.G.: Flexible self-powered piezo-supercapacitor system for wearable electronics. Nanotechnology **29**(32), 325501 (2018)
125. Proto, A., Peter, L., Cerny, M., Penhaker, M., Bibbo, D., Conforto, S., Schmid, M.: Human body energy harvesting solutions for wearable technologies. In: 2018 IEEE 20th International Conference on e-Health Networking, Applications and Services (Healthcom), pp. 1–5, 17–20 Sept 2018 (2018)
126. Starliper, N., Mohammadzadeh, F., Songkakul, T., Hernandez, M., Bozkurt, A., Lobaton, E.: Activity-aware wearable system for power-efficient prediction of physiological responses. Sensors **19**(3), 441 (2019)
127. Khan, M.B., Kim, D.H., Han, J.H., Saif, H., Lee, H., Lee, Y., Kim, M., Jang, E., Hong, S.K., Joe, D.J., Lee, T.-I., Kim, T.-S., Lee, K.J., Lee, Y.: Performance improvement of flexible piezoelectric energy harvester for irregular human motion with energy extraction enhancement circuit. Nano Energy **58**, 211–219 (2019)
128. Elahi, H., Eugeni, M., Gaudenzi, P.: A review on mechanisms for piezoelectric-based energy harvesters. Energies **11**(7), 1850 (2018)
129. Izadgoshasb, I., Lim, Y.Y., Lake, N., Tang, L., Padilla, R.V., Kashiwao, T.: Optimizing orientation of piezoelectric cantilever beam for harvesting energy from human walking. Energy Convers. Manage. **161**, 66–73 (2018)
130. Jeong, S.Y., Hwang, W.S., Cho, J.Y., Jeong, J.C., Ahn, J.H., Kim, K.B., Hong, S.D., Song, G.J., Jeon, D.H., Sung, T.H.: Piezoelectric device operating as sensor and harvester to drive switching circuit in LED shoes. Energy **177**, 87–93 (2019)
131. Chou, X., Zhu, J., Qian, S., Niu, X., Qian, J., Hou, X., Mu, J., Geng, W., Cho, J., He, J., Xue, C.: All-in-one filler-elastomer-based high-performance stretchable piezoelectric nanogenerator for kinetic energy harvesting and self-powered motion monitoring. Nano Energy **53**, 550–558 (2018)
132. Ye, S., Cheng, C., Chen, X., Chen, X., Shao, J., Zhang, J., Hu, H., Tian, H., Li, X., Ma, L., Jia, W.: High-performance piezoelectric nanogenerator based on microstructured P(VDF-TrFE)/BNNTs composite for energy harvesting and radiation protection in space. Nano Energy **60**, 701–714 (2019)
133. Balpande, S.S., Kalambe, J., Pande, R.S.: Vibration energy harvester driven wearable biomedical diagnostic system. In: 2018 IEEE 13th Annual International Conference on Nano/Micro Engineered and Molecular Systems (NEMS), pp. 448–451. IEEE (2018)
134. Huynh, T.P., Haick, H.: Autonomous flexible sensors for health monitoring. Adv. Mater. **30**(50), 1802337 (2018)
135. Kim, M.-O., Pyo, S., Oh, Y., Kang, Y., Cho, K.-H., Choi, J., Kim, J.: Flexible and multi-directional piezoelectric energy harvester for self-powered human motion sensor. Smart Mater. Struct. **27**(3), 035001 (2018)
136. Yan, C., Wang, J., Kang, W., Cui, M., Wang, X., Foo, C.Y., Chee, K.J., Lee, P.S.: Highly stretchable piezoresistive graphene–nanocellulose nanopaper for strain sensors. Adv. Mater. **26**(13), 2022–2027 (2014)
137. Yang, J.-H., Cho, H.-S., Park, S.-H., Song, S.-H., Yun, K.-S., Lee, J.H.: Effect of garment design on piezoelectricity harvesting from joint movement. Smart Mater. Struct. **25**(3), 035012 (2016)
138. Jung, W.-S., Lee, M.-J., Kang, M.-G., Moon, H.G., Yoon, S.-J., Baek, S.-H., Kang, C.-Y.: Powerful curved piezoelectric generator for wearable applications. Nano Energy **13**, 174–181 (2015)
139. Zhang, Y., Wang, T., Luo, A., Hu, Y., Li, X., Wang, F.: Micro electrostatic energy harvester with both broad bandwidth and high normalized power density. Appl. Energy **212**, 362–371 (2018)

140. Tian, Z., He, J., Chen, X., Wen, T., Zhai, C., Zhang, Z., Cho, J., Chou, X., Xue, C.: Core–shell coaxially structured triboelectric nanogenerator for energy harvesting and motion sensing. RSC Adv. **8**(6), 2950–2957 (2018)
141. Jayathilaka, W.A.D.M., Qi, K., Qin, Y., Chinnappan, A., Serrano-García, W., Baskar, C., Wang, H., He, J., Cui, S., Thomas, S.W., Ramakrishna, S.: Significance of nanomaterials in wearables: a review on wearable actuators and sensors. Adv. Mater. **31**(7), 1805921 (2019)
142. Varma, S.J., Sambath Kumar, K., Seal, S., Rajaraman, S., Thomas, J.: Fiber-type solar cells, nanogenerators, batteries, and supercapacitors for wearable applications. Adv. Sci. **5**(9), 1800340 (2018)
143. Zhang, Q., Zhang, Z., Liang, Q., Gao, F., Yi, F., Ma, M., Liao, Q., Kang, Z., Zhang, Y.: Green hybrid power system based on triboelectric nanogenerator for wearable/portable electronics. Nano Energy **55**, 151–163 (2019)
144. Jeon, S.-B., Kim, W.-G., Park, S.-J., Tcho, I.-W., Jin, I.-K., Han, J.-K., Kim, D., Choi, Y.-K.: Self-powered wearable touchpad composed of all commercial fabrics utilizing a crossline array of triboelectric generators. Nano Energy **65**, 103994 (2019)
145. Zhao, X., Han, W., Zhao, C., Wang, S., Kong, F., Ji, X., Li, Z., Shen, X.: Fabrication of Transparent Paper-Based Flexible Thermoelectric Generator for Wearable Energy Harvester Using Modified Distributor Printing Technology. ACS Appl. Mater. Interfaces **11**(10), 10301–10309 (2019)
146. Zhang, X.-S., Su, M., Brugger, J., Kim, B.: Penciling a triboelectric nanogenerator on paper for autonomous power MEMS applications. Nano Energy **33**, 393–401 (2017)
147. Chen, J., Oh, S.K., Nabulsi, N., Johnson, H., Wang, W., Ryou, J.-H.: Biocompatible and sustainable power supply for self-powered wearable and implantable electronics using III-nitride thin-film-based flexible piezoelectric generator. Nano Energy **57**, 670–679 (2019)
148. Hu, D., Yao, M., Fan, Y., Ma, C., Fan, M., Liu, M.: Strategies to achieve high performance piezoelectric nanogenerators. Nano Energy **55**, 288–304 (2019)
149. Park, C.-S., Shin, Y.C., Jo, S.-H., Yoon, H., Choi, W., Youn, B.D., Kim, M.: Two-dimensional octagonal phononic crystals for highly dense piezoelectric energy harvesting. Nano Energy **57**, 327–337 (2019)
150. Mo, X., Zhou, H., Li, W., Xu, Z., Duan, J., Huang, L., Hu, B., Zhou, J.: Piezoelectrets for wearable energy harvesters and sensors. Nano Energy **65**, 104033 (2019)
151. Krebs, F.C.: Fabrication and processing of polymer solar cells: a review of printing and coating techniques. Sol. Energy Mater. Sol. Cells **93**(4), 394–412 (2009)
152. Shepelin, N.A., Lussini, V.C., Fox, P.J., Dicinoski, G.W., Glushenkov, A.M., Shapter, J.G., Ellis, A.V.: 3D printing of poly (vinylidene fluoride-trifluoroethylene): a poling-free technique to manufacture flexible and transparent piezoelectric generators. MRS Commun. **9**(1), 159–164 (2019)
153. Lu, X., Qu, H., Skorobogatiy, M.: Piezoelectric microstructured fibers via drawing of multi-material preforms. In: Energy Harvesting and Storage: Materials, Devices, and Applications VIII, International Society for Optics and Photonics, pp. 106630E (2018)
154. Lu, X., Qu, H., Skorobogatiy, M.: Piezoelectric micro-and nanostructured fibers fabricated from thermoplastic nanocomposites using a fiber drawing technique: comparative study and potential applications. ACS Nano **11**(2), 2103–2114 (2017)
155. Xu, S., Liu, D., Zhang, Q., Fu, Q.: Electric field-induced alignment of nanofibrillated cellulose in thermoplastic polyurethane matrix. Compos. Sci. Technol. **156**, 117–126 (2018)
156. Mokhtari, F., Shamshirsaz, M., Latifi, M.: Investigation of β phase formation in piezoelectric response of electrospun polyvinylidene fluoride nanofibers: LiCl additive and increasing fibers tension. Polym. Eng. Sci. **56**(1), 61 (2016)
157. Zhou, Y., He, J., Wang, H., Qi, K., Nan, N., You, X., Shao, W., Wang, L., Ding, B., Cui, S.: Highly sensitive, self-powered and wearable electronic skin based on pressure-sensitive nanofiber woven fabric sensor. Sci. Rep. **7**(1), 12949 (2017)
158. Vassiliadis, S.G., Matsouka, D.: Piezoelectricity: Organic and Inorganic Materials and Applications. BoD–Books on Demand (2018)

159. Sappati, K.K., Bhadra, S.: Piezoelectric polymer and paper substrates: a review. Sensors **18**(11) (2018)
160. Wan, C., Bowen, C.R.: Multiscale-structuring of polyvinylidene fluoride for energy harvesting: the impact of molecular-, micro- and macro-structure. J. Mater. Chem. A **5**(7), 3091–3128 (2017)
161. Ramadan, K.S., Sameoto, D., Evoy, S.: A review of piezoelectric polymers as functional materials for electromechanical transducers. Smart Mater. Struct. **23**(3), 033001 (2014)
162. Jeong, C.K., Cho, S.B., Han, J.H., Park, D.Y., Yang, S., Park, K.-I., Ryu, J., Sohn, H., Chung, Y.-C., Lee, K.J.: Flexible highly-effective energy harvester via crystallographic and computational control of nanointerfacial morphotropic piezoelectric thin film. Nano Res. **10**(2), 437–455 (2017)
163. Batra, A.K., Aggarwal, M.D.: Pyroelectric Materials: Infrared Detectors, Particle Accelerators and Energy Harvesters. SPIE Press Bellingham, Washington, USA (2013)
164. Maiti, T., Saxena, M., Roy, P.: Double perovskite ($Sr_2B'B''O_6$) oxides for high-temperature thermoelectric power generation—a review. J. Mater. Res. **34**(1), 107–125 (2019)
165. Bowen, C.R., Kim, H.A., Weaver, P.M., Dunn, S.: Piezoelectric and ferroelectric materials and structures for energy harvesting applications. Energy Environ. Sci. **7**(1), 25–44 (2014)
166. Uchino, K.: Piezoelectric energy harvesting systems—essentials to successful developments. Energ. Technol. **6**(5), 829–848 (2018)
167. Cao, J., Ling, M., Inman, D.J., Lin, J.: Generalized constitutive equations for piezo-actuated compliant mechanism. Smart Mater. Struct. **25**(9), 095005 (2016)
168. Deutz, D.B., Pascoe, J.-A., Schelen, B., Van Der Zwaag, S., De Leeuw, D.M., Groen, P.: Analysis and experimental validation of the figure of merit for piezoelectric energy harvesters. Mater. Horiz. **5**(3), 444–453 (2018)
169. Xu, R., Kim, S.: Figures of merits of piezoelectric materials in energy harvesters. In: Proceedings of the PowerMEMS, pp. 464–467 (2012)
170. Roscow, J.I., Pearce, H., Khanbareh, H., Kar-Narayan, S., Bowen, C.R.: Modified energy harvesting figures of merit for stress- and strain-driven piezoelectric systems. Eur. Phys. J. Spec. Top. **228**(7), 1537–1554 (2019)
171. Uchino, K.: Advanced Piezoelectric Materials: Science and Technology. Woodhead Publishing (2017)
172. Kim, H., Fernando, T., Li, M., Lin, Y., Tseng, T.-L.B.: Fabrication and characterization of 3D printed $BaTiO_3$/PVDF nanocomposites. J. Compos. Mater. **52**(2), 197–206 (2018)
173. Nawir, N.A.A., Basari, A.A., Saat, M.S.M., Yan, N.X., Hashimoto, S.: A review on piezoelectric energy harvester and its power conditioning circuit. ARPN J. (2018)
174. Stewart, M., Battrick, W., Cain, M.: Measuring Piezoelectric d_{33} Coefficents Using the Direct Method (2001)
175. Wang, Z., Miao, J.: Critical electrode size in measurement of d_{33} coefficient of films via spatial distribution of piezoelectric displacement. J. Phys. D Appl. Phys. **41**(3), 035306 (2008)
176. Kim, S.-B., Park, J.-H., Kim, S.-H., Ahn, H., Wikle, H.C., Kim, D.-J.: Modeling and evaluation of d_{33} mode piezoelectric energy harvesters. J. Micromech. Microeng. **22**(10), 105013 (2012)
177. Ani Melfa Roji, M., Jiji, G., Ajith Bosco Raj, T.: A retrospect on the role of piezoelectric nanogenerators in the development of the green world. RSC Adv. **7**(53), 33642–33670 (2017)
178. Liu, Y., Wang, Q.: Ferroelectric polymers exhibiting negative longitudinal piezoelectric coefficient: progress and prospects. Adv. Sci. **7**(6), 1902468 (2020)
179. Satthiyaraju, M., Ramesh, T.: Effect of annealing treatment on PVDF nanofibers for mechanical energy harvesting applications. Mater. Res. Express **6**(10), 105366 (2019)
180. Liu, X., Deng, M., Wang, X.: Nanoscale domain imaging and local piezoelectric coefficient d_{33} studies of single piezoelectric polymeric nanofibers. Mater. Lett. **189**, 66–69 (2017)
181. Szewczyk, P.K., Gradys, A., Kim, S.K., Persano, L., Marzec, M., Kryshtal, A., Busolo, T., Toncelli, A., Pisignano, D., Bernasik, A., Kar-Narayan, S., Sajkiewicz, P., Stachewicz, U.: Enhanced piezoelectricity of electrospun polyvinylidene fluoride fibers for energy harvesting. ACS Appl. Mater. Interfaces **12**(11), 13575–13583 (2020)

182. Ponraj, B., Bhimireddi, R., Varma, K.B.R.: Effect of nano- and micron-sized $K_{0.5}Na_{0.5}NbO_3$ fillers on the dielectric and piezoelectric properties of PVDF composites. J. Adv. Ceram. **5**(4), 308–320 (2016)
183. Ahmed, R., Mir, F., Banerjee, S.: A review on energy harvesting approaches for renewable energies from ambient vibrations and acoustic waves using piezoelectricity. Smart Mater. Struct. **26**(8), 085031 (2017)
184. Yuan, H., Lei, T., Qin, Y., Yang, R.: Flexible electronic skins based on piezoelectric nanogenerators and piezotronics. Nano Energy **59**, 84–90 (2019)
185. Han, B., Yu, X., Ou, J.: Challenges of self-sensing concrete. In: Han, B., Yu, X., Ou, J. (eds) Self-Sensing Concrete in Smart Structures, pp. 361–376. Butterworth-Heinemann (2014) (Chapter 11)
186. Invernizzi, F., Dulio, S., Patrini, M., Guizzetti, G., Mustarelli, P.: Energy harvesting from human motion: materials and techniques. Chem. Soc. Rev. **45**(20), 5455–5473 (2016)
187. Tian, Y., Li, G., Yi, Z., Liu, J., Yang, B.: A low-frequency MEMS piezoelectric energy harvester with a rectangular hole based on bulk PZT film. J. Phys. Chem. Solids **117**, 21–27 (2018)
188. Jain, A., Prashanth, K.J., Sharma, A.K., Jain, A., Rashmi, P.N.: Dielectric and piezoelectric properties of PVDF/PZT composites: a review. Polym. Eng. Sci. **55**(7), 1589–1616 (2015)
189. Sappati, K.K., Bhadra, S.: Piezoelectric polymer and paper substrates: a review. Sensors **18**(11), 3605 (2018)
190. Bae, J.-H., Chang, S.-H.: PVDF-based ferroelectric polymers and dielectric elastomers for sensor and actuator applications: a review. Funct. Compos. Struct. **1**(1), 012003 (2019)
191. Fu, J., Hou, Y., Zheng, M., Zhu, M.: Dielectric and energy harvesting properties of $FeTiNbO_6$/PVDF composites with reinforced sandwich structure. Polym. Compos. **40**(S1), E570–E578 (2019)
192. Oh, W.J., Lim, H.S., Won, J.S., Lee, S.G.: Preparation of PVDF/PAR composites with piezoelectric properties by post-treatment. Polymers **10**(12), 1333 (2018)
193. Maruccio, C., Quaranta, G., Lorenzis, L.D., Monti, G.: Energy harvesting from electrospun piezoelectric nanofibers for structural health monitoring of a cable-stayed bridge. Smart Mater. Struct. **25**(8), 085040 (2016)
194. Kim, K., Ha, M., Choi, B., Joo, S.H., Kang, H.S., Park, J.H., Gu, B., Park, C., Park, C., Kim, J., Kwak, S.K., Ko, H., Jin, J., Kang, S.J.: Biodegradable, electro-active chitin nanofiber films for flexible piezoelectric transducers. Nano Energy **48**, 275–283 (2018)
195. Kim, M., Wu, Y.S., Kan, E.C., Fan, J.: Breathable and flexible piezoelectric ZnO@PVDF fibrous nanogenerator for wearable applications. Polymers **10**(7), 745 (2018)
196. Kim, S.-R., Yoo, J.-H., Cho, Y.S., Park, J.-W.: Flexible piezoelectric energy generators based on P(VDF-TrFE) nanofibers. Mater. Res. Express **6**(8), 086311 (2019)
197. Ren, X., Meng, N., Yan, H., Bilotti, E., Reece, M.J.: Remarkably enhanced polarisability and breakdown strength in PVDF-based interactive polymer blends for advanced energy storage applications. Polymer **168**, 246–254 (2019)
198. Martins, P., Lopes, A.C., Lanceros-Mendez, S.: Electroactive phases of poly(vinylidene fluoride): determination, processing and applications. Prog. Polym. Sci. **39**(4), 683–706 (2014)
199. Jeong, K., Kim, D.H., Chung, Y.S., Hwang, S.K., Hwang, H.Y., Kim, S.S.: Effect of processing parameters of the continuous wet spinning system on the crystal phase of PVDF fibers. J. Appl. Polym. Sci. **135**(3), 45712 (2018)
200. Teka, A., Bairagi, S., Shahadat, M., Joshi, M., Ziauddin Ahammad, S., Wazed Ali, S.: Poly(vinylidene fluoride) (PVDF)/potassium sodium niobate (KNN)–based nanofibrous web: a unique nanogenerator for renewable energy harvesting and investigating the role of KNN nanostructures. Polym. Adv. Technol. **29**(9), 2537–2544 (2018)
201. Kim, D.H., Hwang, H.Y., Kim, S.S.: Fabrication of PVDF film sensors for fatigue damage monitoring of single-lap adhesive joints. Compos. Struct. **207**, 223–231 (2019)

202. Teka, A., Bairagi, S., Shahadat, M., Joshi, M., Ahammad, S.Z., Ali, S.W.: Poly (vinylidene fluoride)(PVDF)/potassium sodium niobate (KNN)–based nanofibrous web: a unique nanogenerator for renewable energy harvesting and investigating the role of KNN nanostructures. Polym. Adv. Technol. **29**(9), 2537–2544 (2018)
203. Wu, L., Jing, M., Liu, Y., Ning, H., Liu, X., Liu, S., Lin, L., Hu, N., Liu, L.: Power generation by PVDF-TrFE/graphene nanocomposite films. Compos. B Eng. **164**, 703–709 (2019)
204. Ma, J., Zhang, Q., Lin, K., Zhou, L., Ni, Z.: Piezoelectric and optoelectronic properties of electrospinning hybrid PVDF and ZnO nanofibers. Mater. Res. Express **5**(3), 035057 (2018)
205. Hofmann, P., Walch, A., Dinkelmann, A., Selvarayan, S.K., Gresser, G.T.: Woven piezoelectric sensors as part of the textile reinforcement of fiber reinforced plastics. Compos. A **116**, 79–86 (2019)
206. Li, Z., Shen, J., Abdalla, I., Yu, J., Ding, B.: Nanofibrous membrane constructed wearable triboelectric nanogenerator for high performance biomechanical energy harvesting. Nano Energy **36**, 341–348 (2017)
207. Li, Z., Zhu, M., Shen, J., Qiu, Q., Yu, J., Ding, B.: All-fiber structured electronic skin with high elasticity and breathability. Adv. Func. Mater. **30**(6), 1908411 (2020)
208. Shen, J., Li, Z., Yu, J., Ding, B.: Humidity-resisting triboelectric nanogenerator for high performance biomechanical energy harvesting. Nano Energy **40**, 282–288 (2017)
209. Mokhtari, F., Foroughi, J., Latifi, M.: Enhancing β crystal phase content in electrospun PVDF nanofibers. In: Energy Harvesting Properties of Electrospun Nanofibers, pp. 5-1–5-28. IOP Publishing (2019)
210. Ruan, L., Yao, X., Chang, Y., Zhou, L., Qin, G., Zhang, X.: Properties and applications of the β phase poly(vinylidene fluoride). Polymers **10**(3), 228 (2018)
211. Karimi, S., Ghaee, A., Barzin, J.: Preparation and characterization of a piezoelectric poly (vinylidene fluoride)/nanohydroxyapatite scaffold capable of naproxen delivery. Eur. Polymer J. **112**, 442–451 (2019)
212. Yu, S.M., Oh, H.J., Hwang, S.-K., Chung, Y.-S., Kim, S.-S.: The Effects of Post-Treatments for Wet Spun PVDF on the Piezoelectric Property, p. 26 (2013)
213. De Neef, A., Samuel, C., Stoclet, G., Rguiti, M., Courtois, C., Dubois, P., Soulestin, J., Raquez, J.-M.: Processing of PVDF-based electroactive/ferroelectric films: importance of PMMA and cooling rate from the melt state on the crystallization of PVDF beta-crystals. Soft Matter **14**(22), 4591–4602 (2018)
214. Divya, S., Hemalatha, J.: Study on the enhancement of ferroelectric β phase in P(VDF-HFP) films under heating and poling conditions. Eur. Polymer J. **88**, 136–147 (2017)
215. Lee, H.J., Hong, T.M., Lim, S.C., Won, J.S., Lee, S.G.: Preparation and characterization of PVDF/PU bicomponent nanofiber by electrospinning. Text. Sci. Eng. **52**(2), 88–96 (2015)
216. Parangusan, H., Ponnamma, D., Al-Maadeed, M.A.A.: Stretchable electrospun PVDF-HFP/Co-ZnO nanofibers as piezoelectric nanogenerators. Sci. Rep. **8**(1), 754 (2018)
217. Naik, R., Somasekhara Rao, T.: Self-powered flexible piezoelectric nanogenerator made of poly (vinylidene fluoride)/Zirconium oxide nanocomposite. Mater. Res. Express **6**(11), 115330 (2019)
218. Dhatarwal, P., Sengwa, R.J.: Polymer compositional ratio-dependent morphology, crystallinity, dielectric dispersion, structural dynamics, and electrical conductivity of PVDF/PEO blend films. Macromol. Res. **27**(10), 1009–1023 (2019)
219. Rajesh, P.S.M., Bodkhe, S., Kamle, S., Verma, V.: Enhancing beta-phase in PVDF through physicochemical modification of cellulose. Electron. Mater. Lett. **10**(1), 315–319 (2014)
220. Chen, J.-J., Li, Y., Zheng, X.-M., He, F.-A., Lam, K.-H.: Enhancement in electroactive crystalline phase and dielectric performance of novel PEG-graphene/PVDF composites. Appl. Surf. Sci. **448**, 320–330 (2018)
221. Bhavanasi, V., Kumar, V., Parida, K., Wang, J., Lee, P.S.: Enhanced piezoelectric energy harvesting performance of flexible PVDF-TrFE bilayer films with graphene oxide. ACS Appl. Mater. Interfaces **8**(1), 521–529 (2016)
222. Lee, J., Lim, S.: Polarization behavior of polyvinylidene fluoride films with the addition of reduced graphene oxide. J. Ind. Eng. Chem. **67**, 478–485 (2018)

223. Mishra, S., Sahoo, R., Unnikrishnan, L., Ramadoss, A., Mohanty, S., Nayak, S.K.: Investigation of the electroactive phase content and dielectric behaviour of mechanically stretched PVDF-GO and PVDF-rGO composites. Mater. Res. Bull. **124**, 110732 (2020)
224. Singh, H.H., Singh, S., Khare, N.: Design of flexible PVDF/$NaNbO_3$/RGO nanogenerator and understanding the role of nanofillers in the output voltage signal. Compos. Sci. Technol. **149**, 127–133 (2017)
225. Kabir, E., Khatun, M., Nasrin, L., Raihan, M.J., Rahman, M.: Pure β-phase formation in polyvinylidene fluoride (PVDF)-carbon nanotube composites. J. Phys. D Appl. Phys. **50**(16), 163002 (2017)
226. Zhu, M., Lou, M., Abdalla, I., Yu, J., Li, Z., Ding, B.: Highly shape adaptive fiber based electronic skin for sensitive joint motion monitoring and tactile sensing. Nano Energy **69**, 104429 (2020)
227. Lou, M., Abdalla, I., Zhu, M., Yu, J., Li, Z., Ding, B.: Hierarchically rough structured and self-powered pressure sensor textile for motion sensing and pulse monitoring. ACS Appl. Mater. Interfaces **12**(1), 1597–1605 (2020)
228. Ren, J.-Y., Zhang, G., Li, Y., Lei, J., Zhu, L., Zhong, G.-J., Li, Z.-M.: Effect of ion-dipole interaction on the formation of polar extended-chain crystals in high pressure-crystallized poly(vinylidene fluoride). Polymer **158**, 204–212 (2018)
229. Yaqoob, U., Kim, H.C.: Enhancement in energy harvesting performances of piezoelectric nanogenerator by sandwiching electrostatic rGO layer between PVDF-BTO layers. In: 2018 IEEE 13th Annual International Conference on Nano/Micro Engineered and Molecular Systems (NEMS), pp. 94–97, 22–26 Apr 2018 (2018)
230. Selvarajan, S., Alluri, N.R., Chandrasekhar, A., Kim, S.-J.: $BaTiO_3$ nanoparticles as biomaterial film for self-powered glucose sensor application. Sens. Actuators B Chem. **234**, 395–403 (2016)
231. Sikarwar, S., Sonker, R.K., Shukla, A., Yadav, B.C.: Synthesis and investigation of cubical shaped barium titanate and its application as opto-electronic humidity sensor. J. Mater. Sci.: Mater. Electron. **29**(15), 12951–12958 (2018)
232. Rusevich, L.L., Zvejnieks, G., Kotomin, E.A., Kržmanc, M.M., Meden, A., Kunej, S., Vlaicu, I.D.: Theoretical and experimental study of (Ba, Sr) TiO_3 perovskite solid solutions and $BaTiO_3$/$SrTiO_3$ heterostructures. J. Phys. Chem. C **123**(4) 2031–2036 (2019)
233. Mayeen, A., Kalarikkal, N.: Development of ceramic-controlled piezoelectric devices for biomedical. In: Fundamental Biomaterials: Ceramics, p. 47 (2018)
234. Choi, H.Y., Jeong, Y.G.: Microstructures and piezoelectric performance of eco-friendly composite films based on nanocellulose and barium titanate nanoparticle. Compos. B Eng. **168**, 58–65 (2019)
235. Jang, S.-M., Yang, S.C.: Highly piezoelectric $BaTiO_3$ nanorod bundle arrays using epitaxially grown TiO_2 nanomaterials. Nanotechnology **29**(23), 235602 (2018)
236. Zhou, B., Li, R., Cai, J., Xu, J., Zhao, Z., Pei, J.: Grain size effect on electric properties of novel $BaTiO_3$/PVDF composite piezoelectric ceramics. Mater. Res. Express **5**(9), 095510 (2018)
237. Shihua, D., Tianxiu, S., Xiaojing, Y., Guanghua, L.: Effect of grain size of $BaTiO_3$ ceramics on dielectric properties. Ferroelectrics **402**(1), 55–59 (2010)
238. Hoshina, T., Takizawa, K., Li, J., Kasama, T., Kakemoto, H., Tsurumi, T.: Domain size effect on dielectric properties of barium titanate ceramics. Jpn. J. Appl. Phys. **47**(9S), 7607 (2008)
239. Dudem, B., Kim, D.H., Bharat, L.K., Yu, J.S.: Highly-flexible piezoelectric nanogenerators with silver nanowires and barium titanate embedded composite films for mechanical energy harvesting. Appl. Energy **230**, 865–874 (2018)
240. Baek, C., Yun, J.H., Wang, H.S., Wang, J.E., Park, H., Park, K.-I., Kim, D.K.: Enhanced output performance of a lead-free nanocomposite generator using $BaTiO_3$ nanoparticles and nanowires filler. Appl. Surf. Sci. **429**, 164–170 (2018)
241. He, J., Wang, H., Su, Z., Guo, Y., Tian, X., Qu, Q., Lin, Y.-L.: Thermal conductivity and electrical insulation of epoxy composites with graphene-SiC nanowires and $BaTiO_3$. Compos. A Appl. Sci. Manuf. **117**, 287–298 (2019)

242. Kim, H.S., Lee, D.W., Kim, D.H., Kong, D.S., Choi, J., Lee, M., Murillo, G., Jung, J.H.: Dominant role of young's modulus for electric power generation in PVDF–$BaTiO_3$ composite-based piezoelectric nanogenerator. Nanomaterials **8**(10), 777 (2018)
243. Shen, L., Gong, L., Chen, S., Zhan, S., Zhang, C., Shao, T.: Improvement of β-phase crystal formation in a $BaTiO_3$-modified PVDF membrane. Plasma Sci. Technol **20**(6), 065510 (2018)
244. Ma, J., Azhar, U., Zong, C., Zhang, Y., Xu, A., Zhai, C., Zhang, L., Zhang, S.: Core-shell structured PVDF@BT nanoparticles for dielectric materials: a novel composite to prove the dependence of dielectric properties on ferroelectric shell. Mater. Des. **164**, 107556 (2019)
245. Chilibon, I., Marat-Mendes, J.N.: Ferroelectric ceramics by sol–gel methods and applications: a review. J. Sol-Gel. Sci. Technol. **64**(3), 571–611 (2012)
246. Lu, J., Liang, X., Yu, W., Hu, S., Shen, S.: Temperature dependence of flexoelectric coefficient for bulk polymer polyvinylidene fluoride. J. Phys. D Appl. Phys. **52**(7), 075302 (2018)
247. Wu, J., $BaTiO_3$-based piezoelectric materials. In: Advances in Lead-Free Piezoelectric Materials, pp. 247–299. Springer Singapore, Singapore (2018)
248. Ma, J.-P., Chen, X.-M., Ouyang, W.-Q., Wang, J., Li, H., Fang, J.-L.: Microstructure, dielectric, and energy storage properties of $BaTiO_3$ ceramics prepared via cold sintering. Ceram. Int. **44**(4), 4436–4441 (2018)
249. Yamada, T., Matsuo, M.: Clothing pressure of knitted fabrics estimated in relation to tensile load under extension and recovery processes by simultaneous measurements. Text. Res. J. **79**(11), 1021–1033 (2009)
250. Kumar, A., Kumar, R., Jain, S.C., Vaish, R.: Vibration induced refrigeration and energy harvesting using piezoelectric materials: a finite element study. RSC Adv. **9**(7), 3918–3926 (2019)
251. Taneja, D.K., Varghese, A., Periasamy, C.: finite element method based performance analysis of piezoelectric materials for nanogenerator applications. In: 2018 Conference on Emerging Devices and Smart Systems (ICEDSS), pp. 102–105. IEEE (2018)
252. Asthana, P., Khanna, G.: Finite-element modeling of piezoelectric energy harvesters using lead-based and lead-free materials for voltage generation. J. Asian Ceram. Soc. **6**(4), 394–400 (2018)
253. Jung, W.-S., Lee, M., Baek, S.-H., Jung, I.K., Yoon, S.-J., Kang, C.-Y.: Structural approaches for enhancing output power of piezoelectric polyvinylidene fluoride generator. Nano Energy **22**, 514–523 (2016)
254. Sun, Y., Chen, J., Li, X., Lu, Y., Zhang, S., Cheng, Z.: Flexible piezoelectric energy harvester/sensor with high voltage output over wide temperature range. Nano Energy **61**, 337–345 (2019)
255. Liu, H., Zhong, J., Lee, C., Lee, S.-W., Lin, L.: A comprehensive review on piezoelectric energy harvesting technology: materials, mechanisms, and applications. Appl. Phys. Rev. **5**(4), 041306 (2018)
256. Song, S., Yun, K.-S.: Design and characterization of scalable woven piezoelectric energy harvester for wearable applications. Smart Mater. Struct. **24**(4), 045008 (2015)
257. Wang, W.-C., Wu, L.-Y., Chen, L.-W., Liu, C.-M.: Acoustic energy harvesting by piezoelectric curved beams in the cavity of a sonic crystal. Smart Mater. Struct. **19**(4), 045016 (2010)
258. Hu, P., Yan, L., Zhao, C., Zhang, Y., Niu, J.: Double-layer structured PVDF nanocomposite film designed for flexible nanogenerator exhibiting enhanced piezoelectric output and mechanical property. Compos. Sci. Technol. **168**, 327–335 (2018)
259. Sankaran, S., Deshmukh, K., Ahamed, M.B., Pasha, S.K.: Recent advances in electromagnetic interference shielding properties of metal and carbon filler reinforced flexible polymer composites: a review. Compos. Part A Appl. Sci. Manufact. (2018)
260. Al-Gharabli, S., Kujawa, J., Mavukkandy, M.O., Agbaje, T.A., Hamad, E.M., Arafat, H.A.: Covalent surface entanglement of polyvinylidene fluoride membranes with carbon nanotubes. Eur. Polymer J. **100**, 153–164 (2018)
261. Zhao, B., Zhao, C., Hamidinejad, M., Wang, C., Li, R., Wang, S., Yasamin, K., Park, C.B.: Incorporating a microcellular structure into PVDF/graphene–nanoplatelet composites to tune their electrical conductivity and electromagnetic interference shielding properties. J. Mater. Chem. C **6**(38), 10292–10300 (2018)

262. Bian, Y., Zhang, Y., Sun, K., Jin, H., Dai, L., Shen, H.: A biomimetic vibration sensor using a symmetric electrodes metal core piezoelectric fiber. J. Intell. Mater. Syst. Struct. **29**(6), 1015–1024 (2018)
263. Sim, H.J., Choi, C., Lee, C.J., Kim, Y.T., Spinks, G.M., Lima, M.D., Baughman, R.H., Kim, S.J.: Flexible, stretchable and weavable piezoelectric fiber. Adv. Eng. Mater. **17**(9), 1270–1275 (2015)
264. Egusa, S., Wang, Z., Chocat, N., Ruff, Z.M., Stolyarov, A.M., Shemuly, D., Sorin, F., Rakich, P.T., Joannopoulos, J.D., Fink, Y.: Multimaterial piezoelectric fibres. Nat. Mater. **9**(8), 643–648 (2010)
265. Li, B., Zhang, F., Guan, S., Zheng, J., Xu, C.: Wearable piezoelectric device assembled by one-step continuous electrospinning. J. Mater. Chem. C **4**(29), 6988–6995 (2016)
266. Atalay, A., Atalay, O., Husain, M.D., Fernando, A., Potluri, P.: Piezofilm yarn sensor-integrated knitted fabric for healthcare applications. J. Ind. Text. **47**(4), 505–521 (2017)
267. Gao, H., Minh, P.T., Wang, H., Minko, S., Locklin, J., Nguyen, T., Sharma, S.: High-performance flexible yarn for wearable piezoelectric nanogenerators. Smart Mater. Struct. **27**(9), 095018 (2018)
268. Xu, S., Qin, Y., Xu, C., Wei, Y., Yang, R., Wang, Z.L.: Self-powered nanowire devices. Nat. Nanotechnol. **5**(5), 366–373 (2010)
269. Ghosh, S.K., Mandal, D.: Synergistically enhanced piezoelectric output in highly aligned 1D polymer nanofibers integrated all-fiber nanogenerator for wearable nano-tactile sensor. Nano Energy **53**, 245–257 (2018)
270. Zeng, W., Tao, X.-M., Chen, S., Shang, S., Chan, H.L.W., Choy, S.H.: Highly durable all-fiber nanogenerator for mechanical energy harvesting. Energy Environ. Sci. **6**(9), 2631–2638 (2013)
271. Nilsson, E., Lund, A., Jonasson, C., Johansson, C., Hagström, B.: Poling and characterization of piezoelectric polymer fibers for use in textile sensors. Sens. Actuators A **201**, 477–486 (2013)
272. Ahn, Y., Song, S., Yun, K.-S.: Woven flexible textile structure for wearable power-generating tactile sensor array. Smart Mater. Struct. **24**(7), 075002 (2015)
273. Karan, S.K., Bera, R., Paria, S., Das, A.K., Maiti, S., Maitra, A., Khatua, B.B.: An approach to design highly durable piezoelectric nanogenerator based on self-poled PVDF/AlO-rGO flexible nanocomposite with high power density and energy conversion efficiency. Adv. Energy Mater. **6**(20), 1601016 (2016)
274. Yang, E., Xu, Z., Chur, L.K., Behroozfar, A., Baniasadi, M., Moreno, S., Huang, J., Gilligan, J., Minary-Jolandan, M.: Nanofibrous smart fabrics from twisted yarns of electrospun piezopolymer. ACS Appl. Mater. Interfaces **9**(28), 24220–24229 (2017)
275. Fan, F.-R., Tian, Z.-Q., Lin Wang, Z.: Flexible triboelectric generator. Nano Energy **1**(2), 328–334 (2012)
276. Kim, W., Pyo, S., Kim, M.-O., Oh, Y., Kwon, D.-S., Kim, J.: Humidity-resistant triboelectric energy harvester using electrospun PVDF/PU nanofibers for flexibility and air permeability. Nanotechnology **30**(27), 275401 (2019)
277. Sim, H.J., Choi, C., Kim, S.H., Kim, K.M., Lee, C.J., Kim, Y.T., Lepró, X., Baughman, R.H., Kim, S.J.: Stretchable triboelectric fiber for self-powered kinematic sensing textile. Sci. Rep. **6**, 35153 (2016)
278. Yaqoob, U., Uddin, A.S.M.I., Chung, G.-S.: A novel tri-layer flexible piezoelectric nanogenerator based on surface-modified graphene and PVDF-$BaTiO_3$ nanocomposites. Appl. Surf. Sci. **405**, 420–426 (2017)
279. Shen, Z.-Y., Luo, W.-Q., Li, Y.-M., Hu, Q.-G., Wang, Z.-M., Gu, X.-Y.: Electrical heterostructure of $Nd_{0.1}Sr_{0.9}TiO_3$ ceramic for energy storage applications. J. Mater. Sci. Mater. Electron. **24**(2), 607–612 (2013)
280. Hao, Y.N., Wang, X.H., O'Brien, S., Lombardi, J., Li, L.T.: Flexible $BaTiO_3$/PVDF gradated multilayer nanocomposite film with enhanced dielectric strength and high energy density. J. Mater. Chem. C **3**(37), 9740–9747 (2015)

281. Yang, Y., Gao, Z.-S., Yang, M., Zheng, M.-S., Wang, D.-R., Zha, J.-W., Wen, Y.-Q., Dang, Z.-M.: Enhanced energy conversion efficiency in the surface modified $BaTiO_3$ nanoparticles/polyurethane nanocomposites for potential dielectric elastomer generators. Nano Energy **59**, 363–371 (2019)
282. Patra, A., Pal, A., Sen, S.: Polyvinylpyrrolidone modified barium zirconate titanate /polyvinylidene fluoride nanocomposites as self-powered sensor. Ceram. Int. **44**(10), 11196–11203 (2018)
283. Lin, Z.-H., Yang, Y., Wu, J.M., Liu, Y., Zhang, F., Wang, Z.L.: $BaTiO_3$ nanotubes-based flexible and transparent nanogenerators. J. Phys. Chem. Lett. **3**(23), 3599–3604 (2012)
284. Song, J., Yang, B., Zeng, W., Peng, Z., Lin, S., Li, J., Tao, X.: Highly flexible, large-area, and facile textile-based hybrid nanogenerator with cascaded piezoelectric and triboelectric units for mechanical energy harvesting. Adv. Mater. Technol. **3**(6), 1800016 (2018)
285. Lu, X., Qu, H., Skorobogatiy, M.: Piezoelectric micro- and nanostructured fibers fabricated from thermoplastic nanocomposites using a fiber drawing technique: comparative study and potential applications. ACS Nano **11**(2), 2103–2114 (2017)
286. Siddiqui, S., Kim, D.-I., Roh, E., Duy, L.T., Trung, T.Q., Nguyen, M.T., Lee, N.-E.: A durable and stable piezoelectric nanogenerator with nanocomposite nanofibers embedded in an elastomer under high loading for a self-powered sensor system. Nano Energy **30**, 434–442 (2016)
287. Nunes-Pereira, J., Sencadas, V., Correia, V., Cardoso, V.F., Han, W., Rocha, J.G., Lanceros-Méndez, S.: Energy harvesting performance of $BaTiO_3$/poly(vinylidene fluoride–trifluoroethylene) spin coated nanocomposites. Compos. B Eng. **72**, 130–136 (2015)
288. Mokhtari, F., Spinks, G.M., Fay, C., Cheng, Z., Raad, R., Xi, J., Foroughi, J.: Wearable electronic textiles from nanostructured piezoelectric fibers. Adv. Mater. Technol. **n/a**(n/a), 1900900
289. Zhao, Y., Liao, Q., Zhang, G., Zhang, Z., Liang, Q., Liao, X., Zhang, Y.: High output piezoelectric nanocomposite generators composed of oriented $BaTiO_3$ NPs@PVDF. Nano Energy **11**, 719–727 (2015)
290. Hu, X., Yan, X., Gong, L., Wang, F., Xu, Y., Feng, L., Zhang, D., Jiang, Y.: Improved piezoelectric sensing performance of P(VDF–TrFE) nanofibers by utilizing BTO nanoparticles and penetrated electrodes. ACS Appl. Mater. Interfaces **11**(7), 7379–7386 (2019)
291. Lee, M., Chen, C.-Y., Wang, S., Cha, S.N., Park, Y.J., Kim, J.M., Chou, L.-J., Wang, Z.L.: A hybrid piezoelectric structure for wearable nanogenerators. Adv. Mater. **24**(13), 1759–1764 (2012)
292. Kim, D.H., Dudem, B., Yu, J.S.: High-performance flexible piezoelectric-assisted triboelectric hybrid nanogenerator via polydimethylsiloxane-encapsulated nanoflower-like ZnO composite films for scavenging energy from daily human activities. ACS Sustain. Chem. Eng. **6**(7), 8525–8535 (2018)
293. Li, X., Lin, Z.-H., Cheng, G., Wen, X., Liu, Y., Niu, S., Wang, Z.L.: 3D fiber-based hybrid nanogenerator for energy harvesting and as a self-powered pressure sensor. ACS Nano **8**(10), 10674–10681 (2014)
294. Chang, C., Tran, V.H., Wang, J., Fuh, Y.-K., Lin, L.: Direct-write piezoelectric polymeric nanogenerator with high energy conversion efficiency. Nano Lett. **10**(2), 726–731 (2010)
295. Liu, Z.H., Pan, C.T., Lin, L.W., Huang, J.C., Ou, Z.Y.: Direct-write PVDF nonwoven fiber fabric energy harvesters via the hollow cylindrical near-field electrospinning process. Smart Mater. Struct. **23**(2), 025003 (2013)
296. Alluri, N.R., Saravanakumar, B., Kim, S.-J.: Flexible, hybrid piezoelectric film ($BaTi_{(1-x)}Zr_xO_3$)/PVDF nanogenerator as a self-powered fluid velocity sensor. ACS Appl. Mater. Interfaces **7**(18), 9831–9840 (2015)
297. Abbasipour, M., Khajavi, R., Yousefi, A.A., Yazdanshenas, M.E., Razaghian, F., Akbarzadeh, A.: Improving piezoelectric and pyroelectric properties of electrospun PVDF nanofibers using nanofillers for energy harvesting application. Polym. Adv. Technol. **30**(2), 279–291 (2019)
298. Eun, Y., Kwon, D.-S., Kim, M.-O., Yoo, I., Sim, J., Ko, H.-J., Cho, K.-H., Kim, J.: A flexible hybrid strain energy harvester using piezoelectric and electrostatic conversion. Smart Mater. Struct. **23**(4), 045040 (2014)

299. Ju, B.-J., Oh, J.-H., Yun, C., Park, C.H.: Development of a superhydrophobic electrospun poly(vinylidene fluoride) web via plasma etching and water immersion for energy harvesting applications. RSC Adv. **8**(50), 28825–28835 (2018)
300. Persano, L., Dagdeviren, C., Su, Y., Zhang, Y., Girardo, S., Pisignano, D., Huang, Y., Rogers, J.A.: High performance piezoelectric devices based on aligned arrays of nanofibers of poly(vinylidenefluoride-co-trifluoroethylene). Nat. Commun. **4**, 1633 (2013)
301. Chen, X., Tian, H., Li, X., Shao, J., Ding, Y., An, N., Zhou, Y.: A high performance P(VDF-TrFE) nanogenerator with self-connected and vertically integrated fibers by patterned EHD pulling. Nanoscale **7**(27), 11536–11544 (2015)
302. Tiwari, S., Gaur, A., Kumar, C., Maiti, P.: Enhanced piezoelectric response in nanoclay induced electrospun PVDF nanofibers for energy harvesting. Energy **171**, 485–492 (2019)
303. Guo, W., Tan, C., Shi, K., Li, J., Wang, X.-X., Sun, B., Huang, X., Long, Y.-Z., Jiang, P.: Wireless piezoelectric devices based on electrospun PVDF/$BaTiO_3$ NW nanocomposite fibers for human motion monitoring. Nanoscale **10**(37), 17751–17760 (2018)
304. Vatansever, D., Hadimani, R., Shah, T., Siores, E.: Characterisation of energy generating polyvinylidene fluoride (PVDF) based piezoelectric filament. Adv. Mater. Res. **410**, 366–369 (2012)
305. Hadimani, R.L., Bayramol, D.V., Sion, N., Shah, T., Qian, L., Shi, S., Siores, E.: Continuous production of piezoelectric PVDF fibre for e-textile applications. Smart Mater. Struct. **22**(7), 075017 (2013)
306. Kumar, C., Gaur, A., Tiwari, S., Biswas, A., Rai, S.K., Maiti, P.: Bio-waste polymer hybrid as induced piezoelectric material with high energy harvesting efficiency. Compos. Commun. **11**, 56–61 (2019)
307. Rasel, M.S., Maharjan, P., Park, J.Y.: Hand clapping inspired integrated multilayer hybrid nanogenerator as a wearable and universal power source for portable electronics. Nano Energy **63**, 103816 (2019)
308. Ji, S.H., Yun, J.S.: Fabrication and characterization of aligned flexible lead-free piezoelectric nanofibers for wearable device applications. Nanomaterials **8**(4), 206 (2018)
309. Hu, Y., Kang, W., Fang, Y., Xie, L., Qiu, L., Jin, T.: Piezoelectric poly(vinylidene fluoride) (PVDF) polymer-based sensor for wrist motion signal detection. Appl. Sci. **8**(5), 836 (2018)
310. Singh, D., Choudhary, A., Garg, A.: Flexible and robust piezoelectric polymer nanocomposites based energy harvesters. ACS Appl. Mater. Interfaces **10**(3), 2793–2800 (2018)
311. You, S., Shi, H., Wu, J., Shan, L., Guo, S., Dong, S.: A flexible, wave-shaped P(VDF-TrFE)/metglas piezoelectric composite for wearable applications. J. Appl. Phys. **120**(23), 234103 (2016)
312. Maity, K., Mandal, D.: All-organic high-performance piezoelectric nanogenerator with multi-layer assembled electrospun nanofiber mats for self-powered multifunctional sensors. ACS Appl. Mater. Interfaces **10**(21), 18257–18269 (2018)
313. Fu, J., Hou, Y., Gao, X., Zheng, M., Zhu, M.: Highly durable piezoelectric energy harvester based on a PVDF flexible nanocomposite filled with oriented $BaTi_2O_5$ nanorods with high power density. Nano Energy **52**, 391–401 (2018)
314. Shaikh, M.O., Huang, Y.-B., Wang, C.-C., Chuang, C.-H.: Wearable woven triboelectric nanogenerator utilizing electrospun PVDF nanofibers for mechanical energy harvesting. Micromachines **10**(7), 438 (2019)
315. Matsouka, D., Vassiliadis, S., Vatansever Bayramol, D., Soin, N., Siores, E.: Investigation of the durability and stability of piezoelectric textile fibres. J. Intell. Mater. Syst. Struct. **28**(5), 663–670 (2017)
316. Qiu, Q., Zhu, M., Li, Z., Qiu, K., Liu, X., Yu, J., Ding, B.: Highly flexible, breathable, tailorable and washable power generation fabrics for wearable electronics. Nano Energy **58**, 750–758 (2019)
317. Dhakras, D., Ogale, S.: High-performance organic-inorganic hybrid piezo-nanogenerator via interface enhanced polarization effects for self-powered electronic systems. Adv. Mater. Interfaces **3**(20), 1600492 (2016)

318. Choi, C., Park, J.W., Kim, K.J., Lee, D.W., de Andrade, M.J., Kim, S.H., Gambhir, S., Spinks, G.M., Baughman, R.H., Kim, S.J.: Weavable asymmetric carbon nanotube yarn supercapacitor for electronic textiles. RSC Adv. **8**(24), 13112–13120 (2018)
319. Yetisen, A.K., Martinez-Hurtado, J.L., Ünal, B., Khademhosseini, A., Butt, H.: Wearables in medicine. Adv. Mater. **30**(33), 1706910 (2018)
320. Wearable Technology Database. Vandrico Inc. (2018)
321. IDC Forecasts Steady Double-Digit Growth for Wearables as New Capabilities and Use Cases Expand the Market Opportunities. International Data Corporation, Framingham, MA (2019)
322. India Smart Wearables Market Finishes Strong with 102,000 Shipment Units in 2018Q3, IDC India Reports. International Data Corporation (2018)
323. Worldwide Wearables Market Grows 7.3% in Q3 2017 as Smart Wearables Rise and Basic Wearables Decline. International Data Corporation, Framingham, MA (2017)
324. Li, Y., He, J., Huang, G., Xie, Z.: Development Status and Trends of Wearable Smart Devices on Wrists, pp. 119–129. Springer International Publishing, Cham (2018)
325. Yetisen, A.K., Qu, H., Manbachi, A., Butt, H., Dokmeci, M.R., Hinestroza, J.P., Skorobogatiy, M., Khademhosseini, A., Yun, S.H.: Nanotechnology in textiles. ACS Nano **10**(3), 3042–3068 (2016)
326. Toprakci, H.A., Ghosh, T.K.: Textile sensors. In: Handbook of Smart Textiles, pp. 357–379 (2015)
327. Cherenack, K., van Pieterson, L.: Smart textiles: challenges and opportunities. J. Appl. Phys. **112**(9), 091301 (2012)
328. Åkerfeldt, M., Lund, A., Walkenström, P.: Textile sensing glove with piezoelectric PVDF fibers and printed electrodes of PEDOT:PSS. Text. Res. J. **85**(17), 1789–1799 (2015)
329. Peters, G.M., Sandin, G., Spak, B.: Environmental prospects for mixed textile recycling in Sweden. ACS Sustain. Chem. Eng. **7**(13), 11682–11690 (2019)
330. Jo, C.-H., Myung, S.-T.: Efficient recycling of valuable resources from discarded lithium-ion batteries. J. Power Sources **426**, 259–265 (2019)
331. Ordoñez, J., Gago, E.J., Girard, A.: Processes and technologies for the recycling and recovery of spent lithium-ion batteries. Renew. Sustain. Energy Rev. **60**, 195–205 (2016)
332. Stadlober, B., Zirkl, M., Irimia-Vladu, M.: Route towards sustainable smart sensors: ferroelectric polyvinylidene fluoride-based materials and their integration in flexible electronics. Chem. Soc. Rev. **48**(6), 1787–1825 (2019)
333. Curry, E.J., Ke, K., Chorsi, M.T., Wrobel, K.S., Miller, A.N., Patel, A., Kim, I., Feng, J., Yue, L., Wu, Q., Kuo, C.-L., Lo, K.W.-H., Laurencin, C.T., Ilies, H., Purohit, P.K., Nguyen, T.D.: Biodegradable piezoelectric force sensor. Proc. Natl. Acad. Sci. **115**(5), 909–914 (2018)
334. Wang, X., Xiao, C., Liu, H., Huang, Q., Chen, M.: Fabrication and properties of recycled poly (vinylidene fluoride) (PVDF) hollow fiber membranes. Desalin. Water Treat. **87**, 82–90 (2017)
335. Sandin, G., Peters, G.M.: Environmental impact of textile reuse and recycling—a review. J. Clean. Prod. **184**, 353–365 (2018)

Chapter 2
Materials and Methods

2.1 Materials

The piezoelectric polymer used for meltspinning was poly(vinylidene fluoride) (PVDF) supplied in powder form by Solvay Soleris (Milan, Italy), under the commercial name Solef 6010. The melt flow index (MFI) of Solef 6010 is 2 g/10 min at a load of 2.16 kg (or 6 g/10 min at a load of 5 kg) at 230 °C. Cubic piezoelectric barium titanate nanoparticles with an average diameter of 50 nm and with 99.9% trace metals basis were purchased from Sigma Aldrich Company (China). N,N-dimethylformamide (DMF, >99.8%, Merck) and acetone (>99.8% Merck Chem) as the solvent were used. Conductive silver paint supplies by SPI supplies from USA. Silver plated polyamide yarn (235/36 dtex 4 ply) was purchased from Shieldex, USA. Silicone rubber (Dragon Skin 10 Platinum Silicone—very fast cure) was purchased from Smooth-ON, USA. Woven and knitted conductive fabric, which are silver plated nylon with weight 80 g m^{-2} purchased from Core Electronics Australia.

2.2 Experimental

2.2.1 *Synthesis of Graphene Oxide*

Graphene Oxide (GO) was synthesized from natural graphite powder through a modified hummer method in two steps to reach improvement in oxidation of graphite. The experimental procedure was carried out under a fume hood and the reaction temperature was controlled precisely by a digital thermometer. The chemically converted graphene (CCG) with 1 mg ml^{-1} concentration dispersed in DMF following the reported method [1, 2]. Briefly, an exfoliated aqueous graphene oxide dispersion (0.05 wt%) was chemically reduced by hydrazine and ammonia at 90 °C in two different steps. The H_2SO_4 (5 wt%) added to the dispersion followed by filtration and drying resulted in the formation of agglomerated CCG powder. To prepare a

F. Mokhtari, *Self-Powered Smart Fabrics for Wearable Technologies*, Springer Theses,
https://doi.org/10.1007/978-3-031-06481-4_2

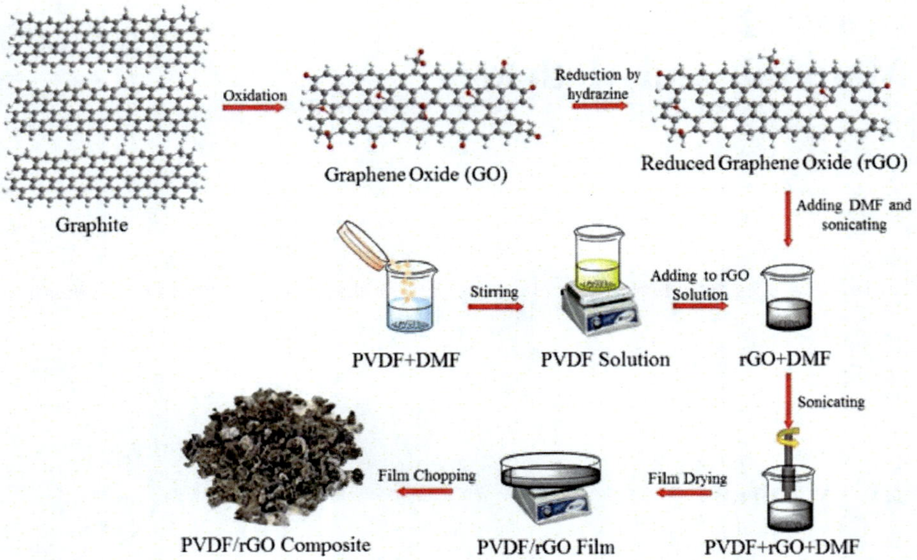

Fig. 2.1 Overall procedure for fabricating PVDF/rGO composite

stable dispersion, the graphene powder was dispersed in DMF using triethylamine by sonication cycles and centrifugation. The direct reduction of graphene oxide in solution helped to avoid rGO aggregation to enhance the rGO dispersion in the PVDF matrix as described in the next section. Moreover, there was no residual reducing agent in the dispersion used to prepare the composites.

2.2.2 Preparation of PVDF/Graphene Composite

To prepare CCG/PVDF composite with 0.5 wt% CCG content, an appropriate amount of CCG dispersion was added to a PVDF solution in DMF (15 w/v%) under constant sonicating and stirring (Fig. 2.1). The mixture was further stirred and sonicated for two hours to ensure homogeneous dispersion of CCG nanosheets into the polymer matrix. After evaporation of the solvent, the resultant composite was chopped into small pieces, washed in ethanol and dried in a vacuum oven at 60 °C for 3 h. The composite was ground into powder manually in a mortar with liquid nitrogen.

2.2.3 Preparation of PVDF/BT Composite

To optimise the ratio of BT nanoparticle in the PVDF polymer matrix for enhancing piezoelectric and mechanical properties of the hybrid PVDF/BT fibers, PVDF/BT nanocomposite films were prepared with different amount of BT nanoparticle. To prepare PVDF/ BT nanocomposite, first 30 g of PVDF powder was dissolved into DMF (200 ml). A clear and transparent solution was obtained upon continuous stirring on heater in water bath at 70 °C overnight. To prepare PVDF/BT nanocomposites solution with different amount of BT nanoparticle wt% (i.e. 5, 10, 15, 20 and 25), the BT nanoparticles were dispersed into DMF (50 ml) using a probe-type ultrasonicate (Fisher Scientific Inc.) at 100 W in 5 s intervals (3 s pulse on, 2 s pulse off) for 60 min under nitrogen flow at 0°C.Then, dispersed BT was added to the as-prepared PVDF (15 wt%) solution and to form a stable suspension, sonicated at 75 W in 5 s intervals (3 s pulse on, 2 s pulse off) for 30 min under nitrogen flow at 0 °C then stirred for 2 h and then put in a vacuum chamber for 5 min to remove air bubbles. The suspensions were cast onto a clean glass plate and after evaporation of the solvent, the film was peeled off. The resultant composite was chopped into small pieces finally to have homogenous mixing, then the chopped film was powdered finely using grinder machine (BARRELL, Scientific, LTD). Polymer pellets and machine's mechanical parts (rotating blades) were cooled down constantly using liquid nitrogen.

2.2.4 Nanostructured Hybrid PVDF/rGO/BT Composite

To prepare PVDF/BT nanocomposite, first 30 g of PVDF powder was dissolved into DMF (200 ml) as described above. To prepare PVDF/BT nanocomposites solution amount 10 wt% of BT nanoparticle was dispersed into DMF (50 ml) using a probe sonicator for 60 min under nitrogen flow at 0 °C then dispersed BT was added to the as-prepared PVDF (10 wt%) solution and to form a stable suspension, sonicated for 30 min under nitrogen flow at 0 °C then stirred for 2 h. Ground powders were prepared from cast films as described above. More details about BT concentration optimization described in chapter four [3]. To prepare the PVDF/rGO/BT nanocomposite, the as-prepared solution of PVDF/rGO/DMF was mixed with the dispersed BT in DMF and sonicated for 30 min under nitrogen flow at −10 °C then stirred for 2 h. The nanocomposite film was made using the same procedure as describe above.

2.2.5 Film Preparation

Two different techniques were studied for making PVDF and PVDF nanocomposite film. These two methods were solution casting and hot press. Although solution casting method for making a PVDF thin film is simple and cost-effective, there is less use in industry because of the difficulty in getting the β phase which has the best piezoelectric response among 4 major types of polymorph. Thus, it would be very interesting if it is possible to obtain polar β phase directly from the solution casting method so further investigations on solution casting was conducted and compared with hot pressing.

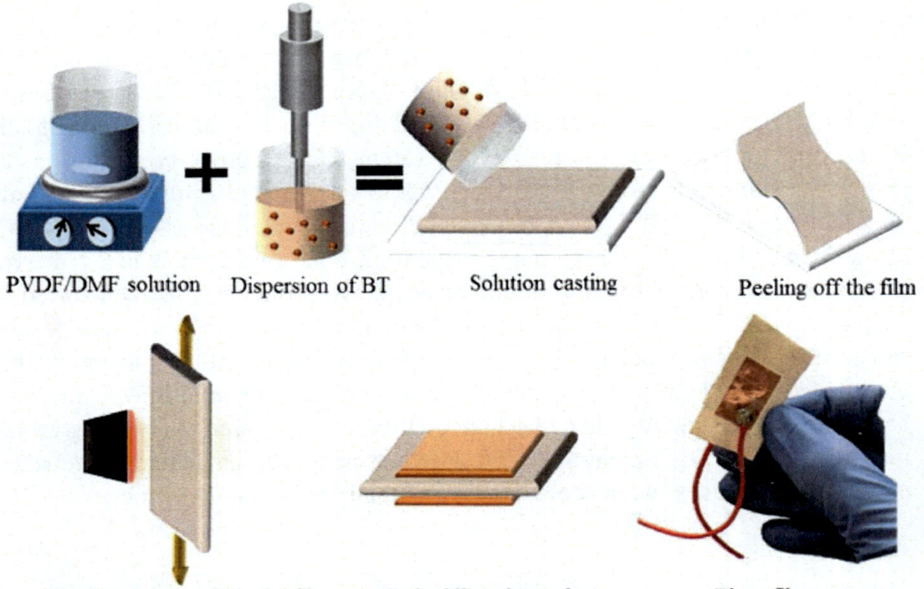

Fig. 2.2 Schematic illustration for the fabrication of PVDF nanocomposite film based flexible pressure sensor

2.2.5.1 Solution Casting Method

The solution casting method associated with appropriate ratio of DMF as well as annealing treatment is proposed as a fast and simple way of producing high-quality PVDF nanocomposite thin film with a large fraction of β phase content. The preparation process of the PVDF/BT nanocomposite film is shown in Fig. 2.2. The prepared nanocomposite as described above was casted onto a clean Petri dish and after evaporation of the solvent, the film was peeled off from the dish. The PVDF/BT nanocomposites film having different concentrations of BT 5, 10, 15, 20 and 25 wt% were obtained from the film casting method. After evaporation of solvent at 90 °C in a vacuum oven a film thickness of 0.4 mm was obtained. To reduce edge effect in the films, the cast PVDF nanocomposite film with an area of 10 cm × 5 cm was trimmed to a final size of 4 cm × 2 cm from the central area of the cast film. After cold drawing process at 80 °C, the film had 13% increase in length. The PVDF films were sandwiched between two cupper tape electrodes under constant pressure to carry out piezoelectric studies.

2.2.5.2 Hot Press Technique

Hot pressing is a common method to fabricate uniform polymer composite films with dense and oriented inner structure. Polymer composite films were prepared by using a hot-press method for the ease of fabrication and its applications. The films were made from the PVDF powder with a Carver hot press (Fig. 2.3a). To create films with higher β crystalline structure, various combinations of heat, pressure, and quenching were used. Then, the optimal hot-pressing conditions were determined. PVDF powder placed between two hot plates of the machine on Teflon plates. Powder samples were compression moulded at 20,000 lbs (800 psi), at 220 °C for 20 min, and then to avoid formation of more α phase, the heating system was turned off and the temperature naturally decreased to room temperature (20 ± 5 °C). The obtained films had the thicknesses of approximately 250 μm and placed between

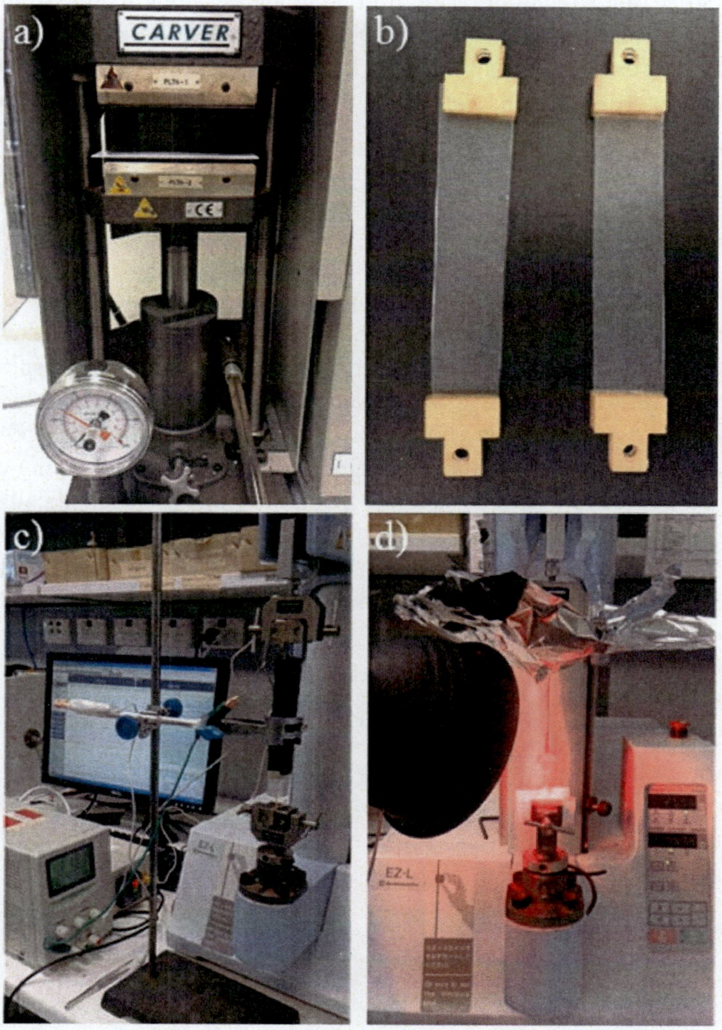

Fig. 2.3 Polymer film preparation process by hot press technique: **a** Hot press apparatus, **b** prepared clear PVDF film between two clamps, **c** stretching PVDF film in a heating chamber and **d** stretching PVDF film using IR lamp

two clamps for cold drawing process (Fig. 2.3b). The PVDF films were stretched by a mechanical tester machine at the constant rate of 10 mm/min up to 100% strain. Heating the polymer film during stretching process was by two different methods: a heating chamber (Fig. 2.3c) or an IR lamp (Fig. 2.3d). Experiments were conducted using an in-house-produced heating chamber. The heating process was operated through Nichrome 80/20 heating wire and driven by an electrical current source. A thermocouple was positioned close to the sample to estimate sample temperature. A slow heating process (3.5 °C min^{-1}) was maintained by using a programmable controller (Electro Chemical Engineering Pty Ltd., Australia) to increase temperature at a uniform rate from 25 to 80 °C. For both methods when the sample temperature reached 80 °C the stretching of the film commenced along its length direction. A necking region and a reduced thickness portion appeared on the film and propagated along the upstretched portions on both sides during stretching. The α to β-phase transformation was reported to initiate at the time of initiation of the necking region.

2.2.6 Meltspinning Process

Melt spinning is the commercially most important fibre spinning process because of its advantages such as high efficiency, energy saving, short and simple flow path. The method is simple: the polymer powder is melted and pumped through a spinneret (die) with one or several holes. The extruded filament is then cooled and stretched in several steps before being collected on a bobbin. Only thermoplastic polymers (polymers that melt) can be used for melt spinning technique. Here melft fiber spinning was carried out using a twin screw extruder (Barrel Scientific Ltd.) as schematically shown in Fig. 2.4.

PVDF powder was heated overnight at a temperature of 70 °C and then fed into the extruder. A single hole spinneret with a diameter of 3 mm was used to produce PVDF

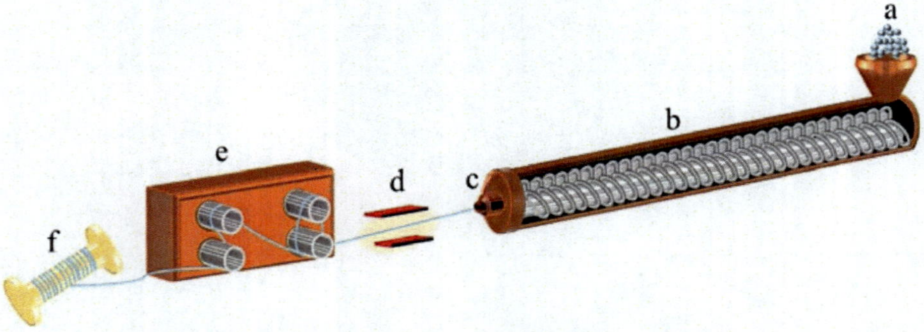

Fig. 2.4 Schematic overview of the melt-spinning set-up used for the production of continuous PVDF fibers: **a** PVDF powder feed; **b** twin screw extruder; **c** single hole 3 mm diameter spinneret (die); **d** heating zone for slow cooling as-spun fibers (80 °C); **e** stretching zone and **f** take up, a spool of melt-spun PVDF piezoelectric fibers

Table 2.1 The extruder heating profile for melt spun fibers

Zoom number	(1)	(2)	(3)	(4)	(5)	(6)	(7)	(8)	(9)
Temperature (°C)	180	190	190	200	210	215	215	220	220

monofilament. The volume of PVDF flowing through the spinneret was controlled to generate a uniform fiber diameter. The temperature profiles along the extruder ranged from 180 to 220 °C over the nine sequential zones (Table 2.1).

To avoid α phase formation due to rapid cooling, the IR lamp was used for slow cooling of the fiber as it extruded from the die. Spinning parameters such as temperature and velocity strongly influence the final properties of melt-spun fibres. The final fiber winding speed of $v_w = 20$ m min^{-1} provides an inline cold drawing ratio $\lambda = v_w/v_t = 10$. The final diameter of the as-prepared PVDF fiber was ~170 μm.

The PVDF/BT nanocomposites fibers were produced through melt-spinning process. As-prepared PVDF/BT nanocomposite films which were prepared through film casting were used to produce the met-spun fibers as explained in Sect. 2.2.5. The melt-spun fiber of PVDF with and without BT were prepared as described above and illustrated in Fig. 2.5. Melt-spun PVDF/BT nanocomposite fibers containing 5, 10 and 20 wt% of the BT were prepared (labelled PVDF/BT_x, where x = 5, 10, 20 is the wt% of BT into the PVDF polymer, (PVDF/BT_5, PVDF/BT_{10} and PVDF/BT_{20}). The final diameter of the stretched PVDF and PVDF/BT nanocomposite fibers was ~170 μm.

2.2.7 *Cold Drawing Process*

After extrusion and drawing of the melt, there will typically be one or more processing steps in the spinning line where the fibres are cold drawn. Cold drawing refers to drawing in the solid state, at a temperature T: $T_g < T < T_m$, where T_g is the glass transition temperature, and T_m is the crystal melting temperature. Prior to cold drawing the fibres exhibit poor mechanical properties, such as low stiffness, high strain to break and high irreversible extensibility. Cold drawing is an irreversible elongation on the macro-scale and, on the micro-scale, an extension and orientation of polymer chains and crystallites along the fibre axis, and the orientation increases with increasing draw ratio. The ratio of the β phase to the α phase was very dependent on the temperature, ratio and speed of drawing. The drawn fibers were prepared through the single stage continuous cold drawing at the feed speed of 3 m min^{-1}. The optimal temperature was set to be 80 °C. Drawing ratio and temperature are two important factors which are studied in this thesis. Figure 2.6 shows set up for cold drawing process using IR lamp as the heater.

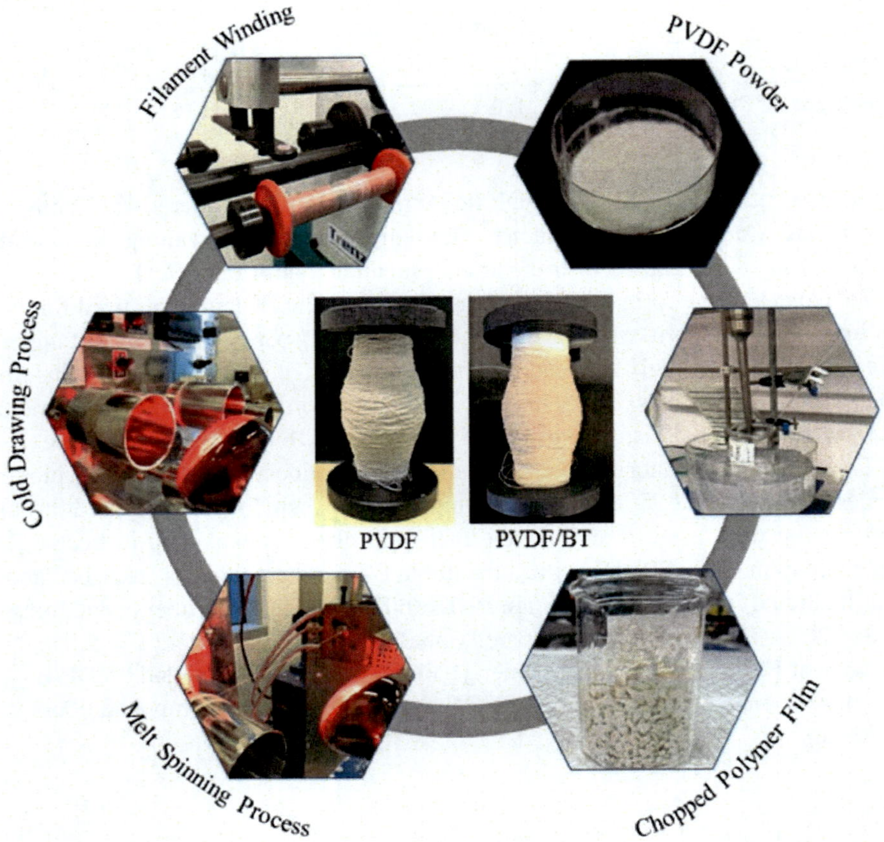

Fig. 2.5 Preparation process of nanocomposite fibers of PVDF/BT

2.2.8 Poling Method

The principle of poling is that a material is exposed to an electrical field and it can be performed in a contact mode and non-contact mode. Contact mode means that there are two electrodes on both sides of the material and the electrodes are connected to a high voltage supply. Non-contact mode, which also can be called corona poling, is when a material is placed between a high potential electrode and grounded counterpart. During poling, the polar crystallites phase of fibers will become aligned with the electric field and this will provide the piezoelectric effect in the fibers. The electrical poling process (during and after drawings) is necessary to improve the performance of the piezoelectric filaments. The poling voltage, time and temperature are effective parameters which depend on the material type and structure and need to be optimised. After application of poling voltage for 30 min the sample was cooled down to the room temperature without disconnecting the external voltage source. The poling process for braid structures and fibers was done by high voltage power

Fig. 2.6 Set up for drawing and heating of PVDF filament

supply of electrospinning machine (MECC. LTD) and poling machine used for films and fabrics. The poling treatment was conducted as follows:

(1) The electrodes embedded on the top and bottom surfaces of specimens to detect the electric charge output.
(2) Install specimens between electrode poles of the poling fixture as shown in Fig. 2.7.
(3) Put the poling fixture into the oil bath, which contains silicone oil for electric insulation, and set 80 °C for the temperature of the oil bath.
(4) Set applied voltage for the poling electric field of the power supply after the temperature of the oil bath reaches 80 °C during preselected times.

For both a single PVDF filament and the triaxial braided PVDF structure poling was carried out using two different methods of electrical contact (fiber axial and radial direction). For the PVDF fibers, a 10 mm section on each end of the PVDF filament was coated with silver paint to act as an electrode. Poling of a 20 mm length of the filament between the electrodes was carried out using different voltages for 10 min with the sample maintained at a temperature of 80 °C. For efficient poling process, the electric field E ($E = V/Z$, where Z is the PVDF filament thickness and V is the applied voltage) should be close to the dielectric strength which depends on the PVDF filament diameter. The dielectric strength is 45–100 MV/m for a PVDF filament of 200 μm diameter which is close to the coercive field for PVDF and its copolymers which range between 50 and 120 MV/m [4]. Care must be taken during this experimental process to avoid flashover or arcing between the electrodes, therefore an insulating rubber tape was placed around the electrode area to serve as an electrical barrier. A small length of one end of the fabricated PVDF braid was modified to expose the core, which served as the inner electrode. The outer sheath

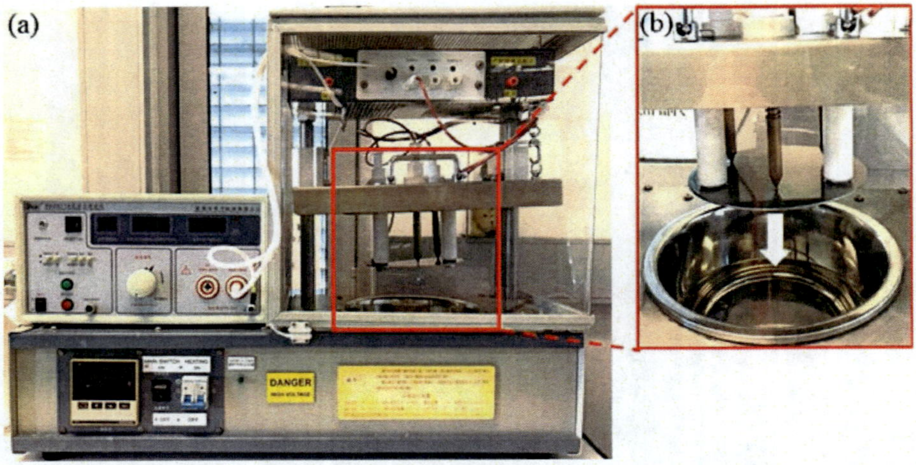

Fig. 2.7 Poling machine to apply electric field to films and fabrics: **a** the poling machine set up with power supply and controlling unit and **b** magnified image for silicon oil bath

was used as the other electrode for the poling process and electrical connection was made at the opposite end of the sample from where the core electrode was connected. The sample was heated to 80 °C for 5 min before the voltage was applied and cooled down to room temperature before the voltage was removed. The applied voltage for the PVDF braid was 25 kV DC.

2.2.9 Fabrication Methods

The as-fabricated piezoelectric filaments are flexible and their strength is high enough to be formed into different structures including braiding, knitting weaving, and coiling which are described in the following sections.

2.2.9.1 Braiding

A triaxial braided piezoelectric PVDF fiber was developed through a multi-step braiding process using a Trenz-Export braiding machine. The process of fabricating the triaxial braided structure is schematically illustrated in Fig. 2.8. A silver coated Nylon (235/36 dtex 4 ply thread) was used in the core as the inner electrode along the length of the fiber with 12 as-prepared PVDF filaments braided around the core to form the piezoelectric layer. Finally, the whole structure was braided a second time with 12 silver coated nylon fibers to serve as the outer electrode. The developed triaxial piezoelectric energy generator can be easily fabricated to unlimited lengths to meet specific energy and power needs. In addition, the triaxial braided structure provides more durability for the piezoelectric energy generator device due to novel

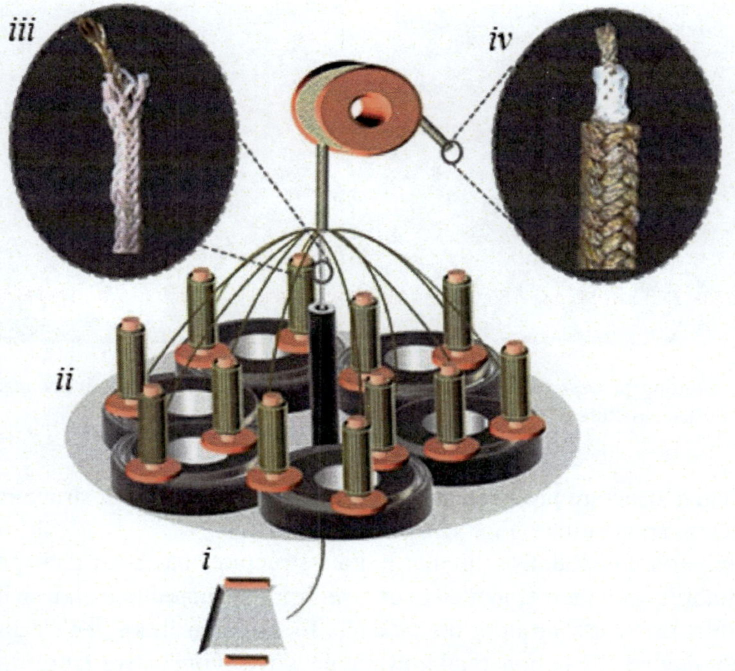

Fig. 2.8 Schematic diagram to illustrate the 3D braiding process: **i** supply spool of the core yarn, **ii** braiding machine with 12 feed spools, **iii** two-layer braided PVDF fibers as the sheath and silver coated nylon the as core, and **iv** fabricated triaxial braided piezoelectric device made by braiding silver coated nylon around the PVDF–silver nylon two-layer fiber

packaging which could protect the PVDF fibers and silver coated nylon electrodes in the device.

2.2.9.2 Circular Knitting

The knitting is the second most common method of textile fabric formation and consists of continuous lengths of yarns which are formed into columns and rows of intermeshed loops. Weft knitted structures are made by loops formed by one continuous end of yarn across the width of the cloth. This technique has the advantage such as: (1) Selective variation of the number of yarns to be knitted in the same fabric, (2) stability and extensibility of the fabric, (3) tuneable porosity and compactness and (4) varying the size of loops. The melt-spun piezoelectric fibers were found to be suitable for use in the knitting machine and could sustain the applied mechanical stress and strain during the knitting process. The fiber diameter to be used for the knitting process was ~170 μm according to the needle size of the machine. Knitted wearable devices had been developed using a Harry Lucas circular knitting machine with the head size of 1/12 in, gauge, 28 and 20 needles (Fig. 2.9a). A feeding tension setting of 10 and pickup tension of 32 was applied to fibers and knitted structures, respectively.

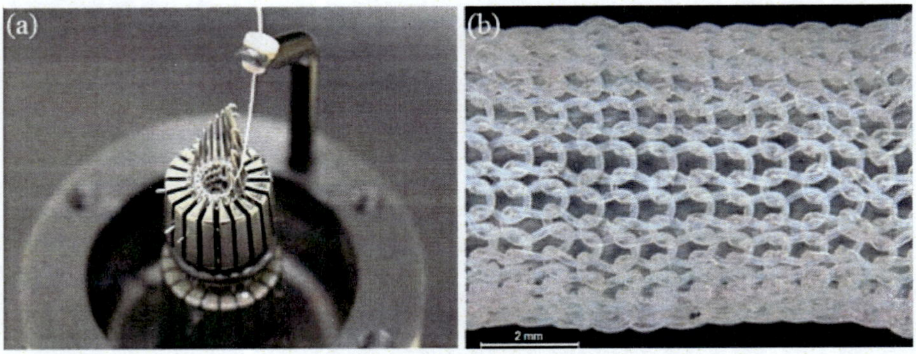

Fig. 2.9 Knitting process for melt-spun piezoelectric fiber: **a** circular knitting machine. **b** 3D circular knitted structure of PVDF fiber

The circular structure provided enough mechanical support and structural integrity for different applications (Fig. 2.9b).

A common complaint of circular knitted structures has been their poor dimensional stability and their tendency to unravel from the unsecured end, so binding-off was performed by sewing along the cut edge. To assemble the knitted wearable energy harvester device, the commercially available woven conductive fabric as inner and outer electrodes with thickness of 80 μm was embedded inside and outside of the knitted structure. The use of the circular knitting would be able to provide more protection against short circuiting as the inner electrode was surrounded by the knitted structure. In addition, stripe electrode bands could be fabricated into the knitted structure as needed by substituting an electrically conductive thread. In this way a wearable knitted device with two knitted electrodes embedded in the structure has been developed.

2.2.9.3 Weaving

The definition of conventional weaving is when two set of yarns/fibres are running orthogonally to one another and interlacing. The yarns running in the width direction in a woven structure are known as weft and the yarns running in the length direction are known as warp. Wearable energy generators and sensors based on PVDF nanocomposites fibers were developed through a weaving process. The preparation was based on plain weave structure in a weaving loom set up (Fig. 2.10a) which is where each weft yarn passes above and below (riser and sinker) the warp yarns repetitively so formed a simple cross pattern (Fig. 2.10b).

In the plain weave the short length of yarn intertwined between warp and weft yarn leads to a fabric with high density and consequently prevents short circuits between the two electrodes. Moreover, the plain weave structure has a homogeneous surface to provide moderately constant electrical properties [5]. To assemble the woven wearable energy harvester device, the commercially available woven conductive

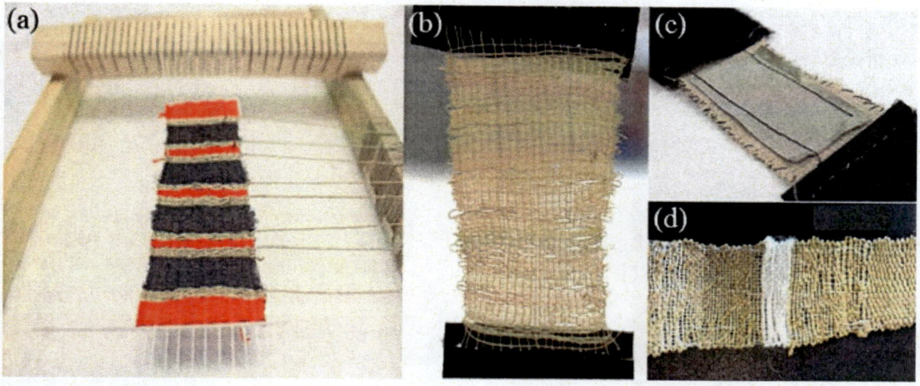

Fig. 2.10 Weaving process for melt-spun piezoelectric fiber: **a** weaving loom set up, **b** woven piezoelectric fiber, **c** placing conductive fabric on both side of woven structure and **d** woven piezoelectric fibers with silver coated Nylon as electrodes

fabric as electrodes with thickness of 80 μm was attached to the top and bottom of the as-prepared fabric using a sewing machine (Fig. 2.10c). Another method to embed electrodes involved weaving silver coated nylon as electrodes on both sides of the woven piezoelectric fiber (Fig. 2.10d).

2.2.9.4 Coiling

To make coils from nanocomposite fibers (Fig. 2.11) as the first step a length of precursor fibers (a_1) were cut off from the fiber spool. The top end of the fiber was joined to the power drill (a_2) by a paper clip (a_3) and a fixed weight (~20 g m) (a_4) hanging on the other end which applied 2 MPa stress to the fiber. The fibers were held in tension but could not rotate due to the applied tension. The weight selection has a key role to have uniform and straight fibers and which also prevent snarl formation prior to coiling.

The precursor fiber was twisted by rotating the powered drill in a clockwise direction (from the top view) to form a "S" twist. The fabricated coil sample (a_5) was mounted on a metallic stand (a_6) stretched for 8% with respect to the initial coil length. while both ends of coil were clamped to prevent twist loss, it was annealed at 120 °C which is above its glass transition (T_g) [6], in an isothermal heating oven. Heating at a temperature over T_g helps to set the twisted shape permanently. After 30 min heating of the coil, it was removed from the oven and left to relax for 2h at room temperature while still clamped. Figure 2.11 schematically shows the process of twist insertion and coil preparation from PVDF fiber.

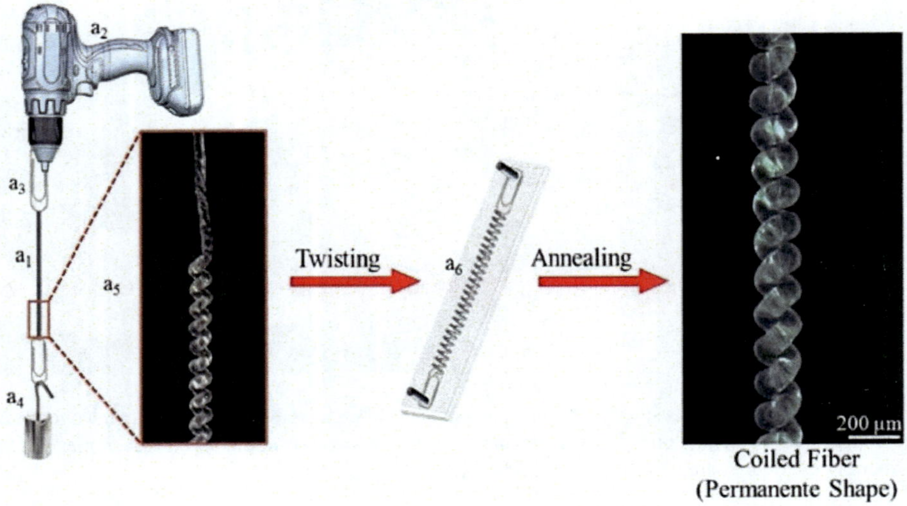

Fig. 2.11 Schematic representation of twist insertion in PVDF nanocomposite fiber using: (a_1) PVDF fiber, (a_2) powered drill, (a_3) metallic paper clip, (a_4) the weight, (a_5) coil formation by induced twist to the fiber (a_6) stretched and clamped coiled fiber

2.2.10 Mechanical Excitation Methods

2.2.10.1 Stretching/Bending Test

The stretching and bending test had been carried out using a Shimadzu EZ mechanical tester to evaluate excitation performance as well as durability of the piezoelectric generator. The stretching of coil and bending strain can be set up for different percentage of strain (10–120%), speed (10–300 mm/min) and cycles at room temperature. Each of the two clamps of the tensile tester was also isolated by insulating tape to insu-late the sample from any extraneous noise generated from the tester mechanism. The Pico Scope clamps were also fixed with tape to the tensile tester clamps to help prevent any movement and noise generation.

2.2.10.2 Arm/Knee Fold-Release Test

To demonstrate a wearable application for energy harvesting, a series of experiments was performed on a human subject as follows. The as-fabricated piezoelectric textile was attached to the arm or knee of myself as a test subject. The electrical properties of the piezoelectric textile were examined by multiple folding-releasing actions of the arm or leg at low frequencies. The generated voltage was transferred to the mobile phone via designing an electrical circuit which was equipped with the Bluetooth technology.

2.2.10.3 Compression and Impact Test

Impact excitation of the piezoelectric sample was performed by two methods. The first method was a dropped weight method (low velocity impact test) to generate an output voltage. The ball weight and size were considered according to the sample size and its sensitivity. Among six different tested balls, a metallic ball (15 mm in diameter and 25 g in weight) was dropped from different heights (5, 10, 15, 20 and 25 cm) through a guide pipe which directed the impact onto the surface of the sample device. To protect samples from potential damage caused by repetitive impacts and also to excite and obtain voltage measurements from large active areas of samples, an aluminium sheet (thickness 5 mm) was placed on samples during the impact tests and the whole system (samples and aluminium sheet) was fixed with clamps to a rigid surface. The electrical response of the piezoelectric textile to the applied stress was recorded by connecting the inner and outer electrodes to an oscilloscope (Pico Scope 4424).

The second method was by an in-house setup that was made for the purpose of comparing all three piezoelectric textile structures and using a periodic impact force. An illustrative graph of the measurement system and assembly details is shown in Fig. 2.12. In order to reduce the interference of triboelectric charges, the samples and impact head were covered with tape. Continuous force impacts were applied at a frequency of 1 Hz. A Nema stepper motor run by Ultimate board was used as the impact power source, and the generated open-circuit voltage and short-circuit current were collected with a Keithley (2612B, USA) system simultaneously.

2.3 Characterization

2.3.1 Mechanical Properties

The mechanical properties of the fibers and films were measured using a Shimadzu tensile tester (EZ-S). Data were recorded by TRAPEZIUMX software on the computer. The samples were mounted between two pneumatic grips. A 10 or 50 N load cell was used in experiments. The Young's modulus was calculated from the slope of the initial part of the stress-strain curve, where the relationship between stress and strain was linear. The reported results for tensile strength, elongation at break and Young's modulus are from 10 times repeat for each type of sample.

2.3.2 Fourier Transform Infrared Spectroscopy (FTIR)

FTIR spectra were obtained using Shimadzu IRPrestige-21 infrared spectrometer equipped with a Pike Technologies Miracle A germanium crystal ATR attachment.

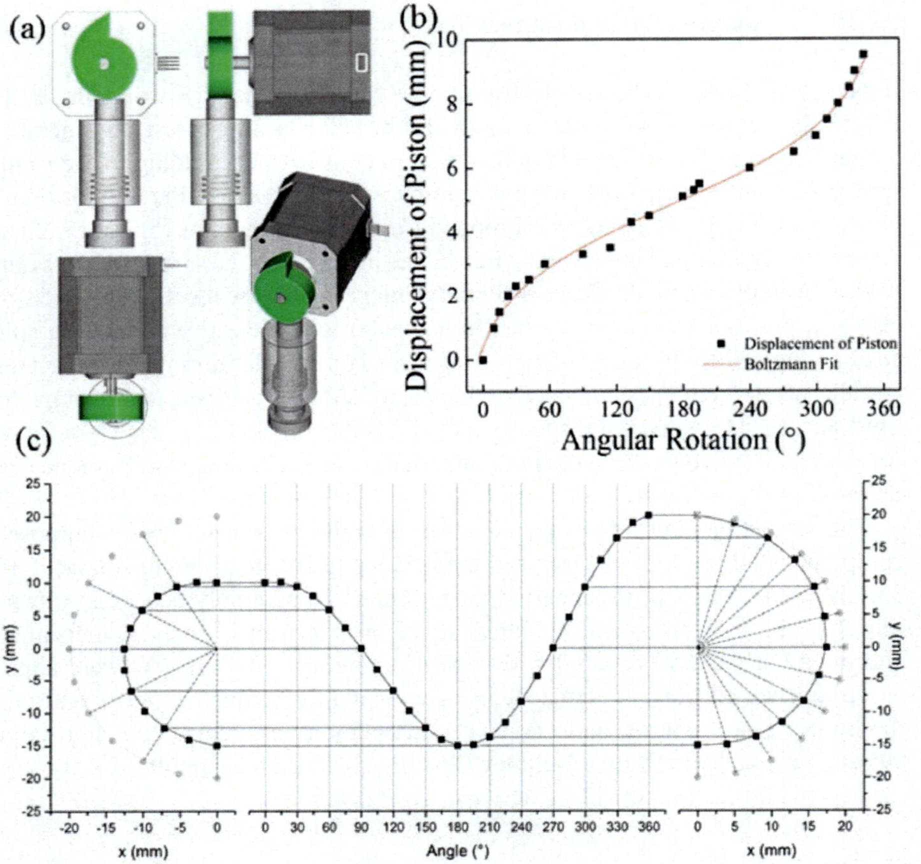

Fig. 2.12 Compression and impact test set up: **a** Multiple views of the in-house setup for continues impact on sample. **b** Measured displacement of the piston with respect to angular rotation. **c** Profile of the designed CAM at different rotational increments

A bunch of filaments which are uniformly arranged alongside each other were used and spectra were recorded over a range of 400–4000 cm^{-1} with 4 cm^{-1} resolution.

2.3.3 X-Ray Diffraction (XRD)

The crystalline structures of samples were analysed by XRD (GBC, MtriX SSD) using Cu Kα radiation ($\lambda = 0.154$ nm), with the generator working at 40 kV and 30 mA. The spectra were obtained between angle of 10° and 80° with typical speeds of 0.5° min^{-1} utilised.

2.3.4 Thermogravimetric Analysis (TGA)

Thermogravimetric Analysis (TGA) was used to determine the decomposition temperature of materials used in the produced fibres. From these results the weight ratio of each filler involved in forming the nanocomposite fibres can be roughly determined. TGA analysis of samples was performed using TG 209 Libra from NETZSCH, Germany. The sample weight of 10 mg was loaded into the TGA crucible and temperature was increased with a ramp rate of 5 °C min^{-1} up to 900 °C under a nitrogen atmosphere.

2.3.5 Differential Scanning Calorimetry (DSC)

DSC analysis was conducted to examine the thermal stability, strength, and behaviour of the fabricated nanocomposite PVDF fibers. The melting temperature (Tm) and melting enthalpy (ΔH_m) of samples were measured with a differential scanning calorimeter (DSC) (TA Instrument) at a heating rate of 5 °C min^{-1} by using a 50 ml/min compressed nitrogen flow as the purging gas. The sample first cooled to 0 °C and then temperature was increased to above melting temperature of the polymer (300 °C).

2.3.6 Dynamic Mechanical Analysis (DMA)

DMA tests were performed using DMA 242 Artemis from NETZSCH, Germany Dynamic Mechanical Analyzer. Storage modulus (E′), loss modulus (E″) and loss factor, tan $\delta = E''/E'$, of 10 mm of PVDF and its nanocomposite fibers were measured in tensile mode following a heating ramp at 3 °C/min from −50 to 150 °C in various frequencies (0.1, 0.2, 0.5, 1, 2, 5, 10 and 20 Hz). The amplitude of the dynamic stress was 0.2 MPa, which was small enough to ensure a linear viscoelastic response from the sample. For each composite, five runs were obtained to get repeatable results since the rheological behaviour depended on the filler dispersion in the polymer matrix.

2.3.7 Scanning Electron Microscopy (SEM)

Surface and cross section morphologies of the samples were examined with a Scanning Electron Microscope (JEOL 6490). Secondary Electron Imaging (SEI) was performed at an accelerating voltage of 10 kV and a probe current setting of 30 mV. Fibres were frozen in liquid nitrogen for 30 min before cutting using a scalpel blade to obtain smooth cross sections for imaging. Samples were coated with 5 nm of Pt using a sputter coater (EDWARDS Auto 306) to aid with imaging and minimise beam heating effects.

2.3.8 Raman Spectroscopy

Raman spectroscopy was performed on samples with GO on Jobin Yvon Horiba HR800 confocal Raman microscope with a 632 nm laser and a 300-line grating under 100× objective to achieve a resolution of $\pm 1.25\ cm^{-1}$.

2.3.9 Filament Diameter Measurement

A stereo microscope (Leica M250A) was used to measure the filament diameter and film thickness by using image analysis tool of Leica software. The reported value is average of 10 times repeat for each measurement.

2.3.10 Electrical Characterization

The electrical properties of both the piezoelectric filaments and fabrics were characterized. In all tests, the open-circuit voltages and short-circuit currents generated by the piezoelectric devices were in-situ recorded by an oscilloscope (Pico Scope4424) with sampling time constant of 0.1 s and Keithley (2612B, USA) with sampling time constant of 10 ms. Also, to minimize the amount of noise, charging capacitors and signal analysis circuits were designed.

2.4 Piezoelectric Characterization

2.4.1 Piezoelectric Force Microscopy (PFM)

Polarization switching in the as-prepared piezoelectric fiber was performed using Piezoresponse Force Microscopy Switching Spectroscopy (SS-PFM) to evaluate the piezoelectric sensitivity. In this method, a direct current voltage (Vdc) was applied as a sequence of pulses, so that the phase and amplitude measurement could be performed in the off-state of pulses. Both the electrostatic force and piezoelectric deformation contribute to the amplitude in the on-state while the electrostatic force is eliminated in the off-state. A significant difference in the off-state compared to the

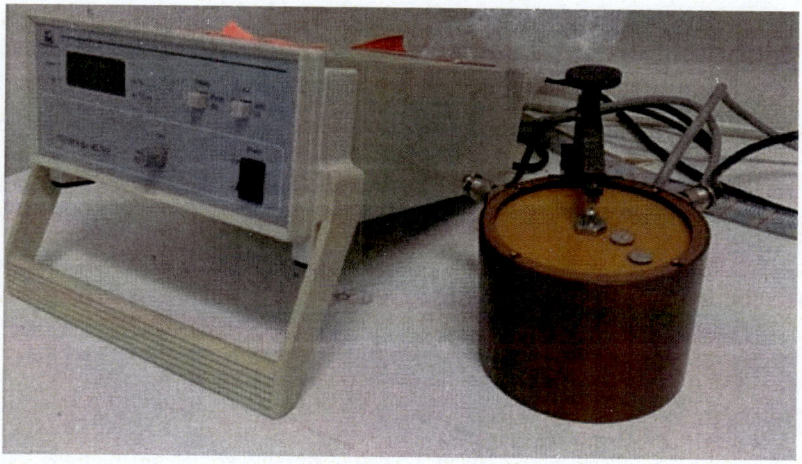

Fig. 2.13 D33 meter machine to measure piezoelectric constant of nanocomposite films

on-state measurements indicated the magnitude of the piezoelectric effect. 24 SS-PFM measurements were carried out using an MFP-3D Atomic Force Microscope (AFM) (Asylum Research, CA).

2.4.2 Piezoelectric Constant Measurement

The piezoelectric coefficient (d_{33} value) of the piezoelectric materials was measured with a quasi-static d_{33} meter (YE2730, APC International. Ltd.) (Fig. 2.13). After calibration of the device with the standard piezo ceramic, the electrodes embedded on both side of the film and placed the film between to jaws of the machine to read the d_{33} base unit of C/N.

2.4.3 Ferroelectric Measurements

Dielectric loss and impedance properties were measured a function of frequency with a precision impedance analyser (4294 A, Agilent Technologies, Inc.) at room temperature with a frequency range from 10^2 to 10^6 Hz. To collect ferroelectric hysteresis loops, a ferroelectric test system (TF2000E, aixACCT, Germany) was used. The samples were loaded into a measurement jig attached to the analyser and placed in contact with a pin. At certain frequencies, a piezoelectric material enters a resonant mode where the electrical input signal excites a mechanical response within the material. These resonant modes are what make a piezoelectric material unique and are not present in dielectric materials. The frequency range used was from 10^2 to 10^6 Hz.

References

1. Sayyar, S., Murray, E., Thompson, B.C., Gambhir, S., Officer, D.L., Wallace, G.G.: Covalently linked biocompatible graphene/polycaprolactone composites for tissue engineering. Carbon **52**, 296–304 (2013)
2. Gambhir, S., Murray, E., Sayyar, S., Wallace, G.G., Officer, D.L.: Anhydrous organic dispersions of highly reduced chemically converted graphene. Carbon **76**, 368–377 (2014)
3. Mokhtari, F., Spinks, G.M., Fay, C., Cheng, Z., Raad, R., Xi, J., Foroughi, J.: Wearable electronic textiles from nanostructured piezoelectric fibers. Adv. Mater. Technol. **n/a**(n/a), 1900900
4. Soin, N., Shah, T.H., Anand, S.C., Geng, J., Pornwannachai, W., Mandal, P., Reid, D., Sharma, S., Hadimani, R.L., Bayramol, D.V.: Novel "3-D spacer" all fibre piezoelectric textiles for energy harvesting applications. Energy Environ. Sci. **7**(5), 1670–1679 (2014)
5. Hofmann, P., Walch, A., Dinkelmann, A., Selvarayan, S.K., Gresser, G.T.: Woven piezoelectric sensors as part of the textile reinforcement of fiber reinforced plastics. Compos. A **116**, 79–86 (2019)
6. McIntyre, J.E.: Synthetic Fibres: Nylon, Polyester, Acrylic, Polyolefin. Taylor & Francis US (2005)

Chapter 3
Triaxial Braided Piezo Fiber Energy Harvester for Self-Powered Wearable Technology

3.1 Introduction

The most common problems associated with current flexible piezoelectric generators and architectures are tedious processing techniques, slow production speed, low output power and lack of comfort [1]. For instance, a novel composite material system and a method for constructing flexible, stretchable and wearable piezoelectric energy-generating fibers have been demonstrated using electrospun polyvinylidene fluoride-*co*-trifluoroethylene (PVDF–TrFE) as the piezoelectric material and carbon nanotubes (CNT) and sliver coated nylon as outer and inner electrodes, respectively [2]. These flexible piezoelectric fibers can be stretched to a tensile strain of 5% and can generate over 50 $\mu W\ cm^{-3}$. Another study demonstrated that a flexible nanogenerator manipulated from a polymer nanocomposite (PVDF–HFP/Co–ZnO) exhibits an output voltage as high as 2.8 V in 50 Hz [3]. A remarkable enhancement of the output voltage (~32 V) of a nanogenerator based on a non-electrically poled cerium (iii) complex containing a PVDF composite film is achieved by simple repeated human finger pressure [4]. In addition, all-fiber piezoelectric fabrics as power generators and energy harvesters have been developed based on 3D spacer technology [1]. Piezoelectric PVDF monofilaments acted as the spacer yarn material interconnected between silver coated polyamide multifilament yarn layers acting as the top and bottom electrodes. This textile structure provided an output power density in the range of 1.10–5.10 $\mu W\ cm^{-2}$ at applied impact pressures in the range of 0.02–0.10 MPa. More recently woven piezoelectric fabrics have been developed using core-sheath melt-spun PVDF/carbon black/polyethylene and silver coated nylon which were able to produce a peak power output exceeding 1 $\mu W\ cm^{-2}$ at an impact pressure of 20 kPa [5]. Although these energy harvesting textiles based on piezoelectric PVDF fibers exhibited stable flexibility and energy harvesting performances, they still suffer from low piezoelectric performance (i.e. power output, durability, and sensitivity) because of the lower interfacial area between the piezoelectric fibers and the outer electrode; in other words, there is poor electrical connection between the piezoelectric fibers and electrodes. To solve this problem, novel fabrication strategies are needed. In this

F. Mokhtari, *Self-Powered Smart Fabrics for Wearable Technologies*, Springer Theses,
https://doi.org/10.1007/978-3-031-06481-4_3

chapter we explore a new type of energy harvesting device developed by braiding melt-spun PVDF piezoelectric and conductive silver coated nylon yarns. The procedure developed for the fabrication is substantially different from conventional energy harvesting PVDF fabrication processes in which two rigid metallic electrodes and piezoelectric fibers are assembled and then fabricated into the energy harvesting generator. In our case, a triaxial structure has been developed based on melt-spun PVDF fibers and silver coated nylon. The as-spun PVDF filaments were first braided around silver coated nylon yarn as a highly flexible inner electrode and then the whole structure was braided a second time with silver coated nylon fibers as an outer electrode. The fabricated device exhibited improved mechanical (i.e. durability, flexibility, and comfort) and piezoelectric properties (i.e. sensitivity and power output) as compared with PVDF fibers. In addition, the fabricated triaxial braided piezoelectric energy harvesting textile provided an efficient and novel way to overcome the stability issues due to the poor fatigue resistance of the metallic electrodes. The triaxial braided piezofiber is strong, lightweight and exceptionally flexible and it is expected to be applicable to wearable devices where high performance is necessary.

3.2 Experimental

3.2.1 Fibers Fabrication and Preparations

Fiber spinning was carried out as described in Sect. 2.2.6. The cold draw process was done as described in Sect. 2.2.7. During stretching, filament whitening was observed, indicating the onset of a phase transformation in the fiber material. The poling process was carried for the PVDF fibers and the triaxial braided PVDF structure was carried out as explained in Sect. 2.2.8. Poling was carried out using voltages of 9, 13, 17 or 20 kV. In addition, to evaluate the effect of the poling process on the piezoelectric response, four different lengths of the as-prepared PVDF filaments were subjected to different applied voltages along their lengths. Table 3.1 shows the length and applied voltage for poling for each of the PVDF filaments. For each sample tested the applied voltage was increased until a spark was observed.

Table 3.1 Different PVDF filament lengths and their applied voltage for the poling process

PVDF filament length (mm)	Applied voltage (kV)
10	9
20	13
40	17
80	20

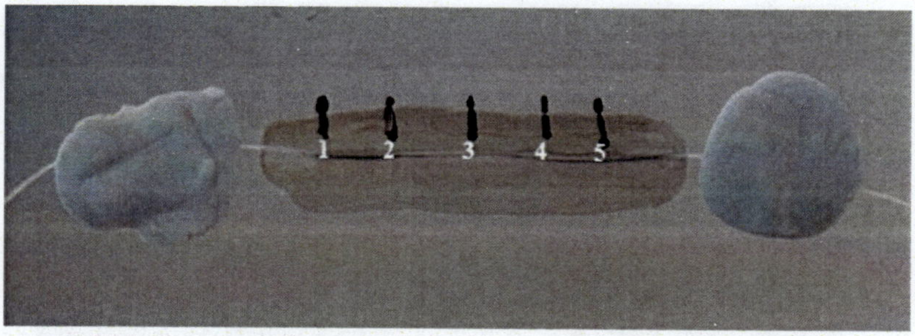

Fig. 3.1 PVDF filament mounted by the silver paste for PFM test

3.2.2 *PFM Test of Single PVDF Fiber*

Polarization switching in the as-prepared PVDF filament was performed as described in Sect. 2.4.1. A small section of the as-spun PVDF fiber was attached to a glass slide using silver paste. A Pt/Ir coated silicon tip (Type: EFM, Nanoworld) with a typical resonance frequency of 75 kHz and a force constant of 2.8 N m^{-1} was used. Scans were performed at various positions along the length of each fiber to find surfaces suitable for PFM testing. The conductive tip and the silver paste were used as top and bottom electrodes to apply a DC voltage to the sample. The conductive tip was triggered with an AC voltage of 500 mV. During the experiments, the vertical piezoresponse was recorded and PFM amplitude-phase scans were performed with switching the DC amplitude up to ± 25 V. The PFM applied voltage is to switch the local dipole underneath the AFM tip. The PFM applied electric field in this thesis agrees with those previously reported in previous researches [6–8]. SS-PFM testing was carried out on PVDF filaments of different lengths (1, 2, 4, and 8 cm). The as-prepared PVDF fiber of 10 mm in length was mounted on a glass slide using silver paste and 5 evenly spaced positions were defined along the fiber length with a marker (Fig. 3.1). During testing, the probe was pressed firmly onto the specified testing point on the sample with a static pressing force of about 40 nN. A large driving voltage of 10 V was then applied in order to generate large amplitude responses.

3.2.3 *Fabrication of Triaxial Braided PVDF Fibers*

A triaxial braided piezoelectric PVDF fiber was developed as described in Sect. 2.2.9. (Fig. 3.2). The developed triaxial piezoelectric energy generator can be easily fabricated with unlimited lengths to meet specific energy and power needs. In addition, the triaxial braided structure provides more durability for the piezoelectric energy generator device due to novel packaging which could protect the PVDF fibers and silver coated nylon electrodes in the device.

Fig. 3.2 Optical photograph of the as-developed triaxial braided piezo generator: (i) silver coated nylon as inner and outer electrodes and (ii) braided PVDF fibers as the intermediate layer. The structure had a total diameter of 2.55 mm consisting of a silver coated nylon core (0.6 mm), an intermediate PVDF layer with a thickness of 1.37 mm, and outer sheath of silver coated nylon braid with a thickness of 0.58 mm

3.2.4 *Sample Excitation Method*

The excitation of the as-prepared piezoelectric PVDF samples was carried out as explained in Sect. 2.2.10 using both impact and bending methods. Bending was induced by compressive loading of the sample until buckling occurred (Fig. 3.3). The compression strain was set to 50% of the sample length (20 mm) with a speed of 300 mm min^{-1} over 1000 bending cycles (Fig. 3.3). Each of the two clamps of the tensile tester was also isolated by insulating tape to insulate the sample from any extraneous noise generated from the tester mechanism. Picoscope clamps were also fixed with tape to the tensile tester clamps to help prevent any movement.

3.3 Results and Discussion

The melt-spun piezoelectric fiber and novel manufacturing process have been developed to create an energy harvesting generator based on the piezoelectric textile. The piezoelectric textile was produced using low cost materials and manufactured with readily scalable textile manufacturing methods.

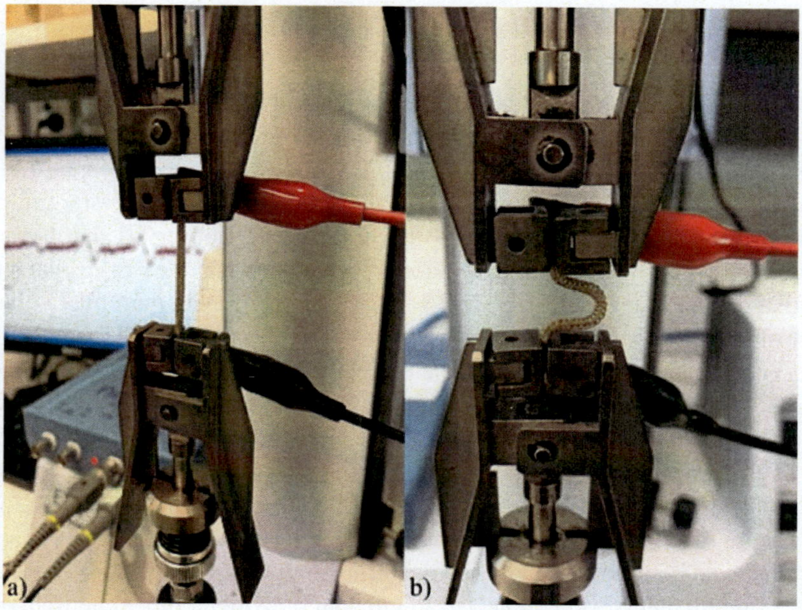

Fig. 3.3 A photograph of the mechanical bending setup: the as-prepared braided piezoelectric device **a** before and **b** after subjected to 50% strain

3.3.1 As-Prepared PVDF Fiber

SEM micrographs of the as-spun PVDF fibers are shown in Fig. 3.4. It shows circular cross-section and a dense internal structure (Fig. 3.4a) and a smooth surface morphology (Fig. 3.4b). The high magnification images (inset) of the surface and cross-section of the PVDF fiber confirmed that there are no voids in the as-prepared PVDF fiber [9].

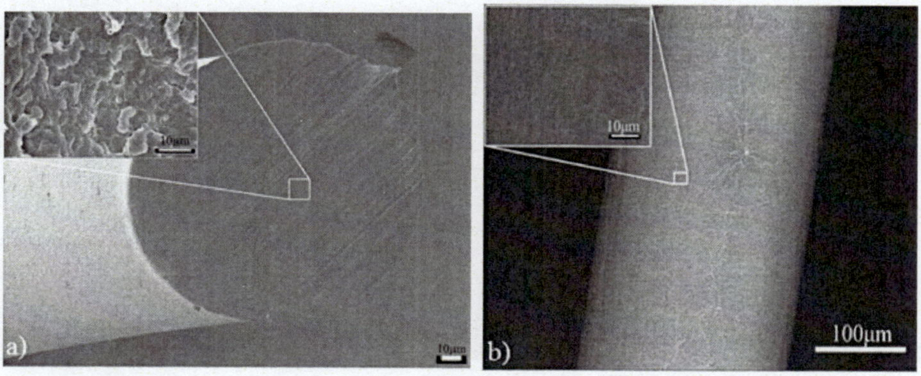

Fig. 3.4 SEM images of the as-prepared PVDF fiber: **a** cross section of the stretched melt-spun PVDF fiber at low and high magnification (inset), **b** PVDF fiber along the length

3.3.2 PVDF Characterisation

The predominant crystalline phase of the as-prepared PVDF fiber following different post-treatment processes (i.e. drawing and poling process) was investigated using FTIR, DSC, and X-ray diffraction. The FTIR spectra of the melt-spun filaments, stretched and poled PVDF fibers in triaxial braid structure and the starting powder are shown in Fig. 3.5 and Table 3.2. It can be clearly observed that the starting powder consists primarily of the α-phase, as is evident from the strong characteristic absorbance bands at 760, 796 and 974 cm^{-1}. For poled melt-spun fibers, the emergence of peaks at 840 (CH_2 rocking), 1276 (trans band) and 1430 (CH_2 bending) cm^{-1} corresponding to the β phase can be clearly observed, along with a corresponding decrease in the peak at 760 cm^{-1}. Similar results have been observed by various authors where it was shown that both the stretching temperatures (80–100 °C), as well as the drawing ratios, influence the β-phase content. It has been suggested by Lund et al. [10] that an increase in the drawing ratio leads to a higher conversion from the α-phase to the β-phase with a strong correlation between the drawing speeds and the conversion between the phases [10]. During melt extrusion of the PVDF monofilaments, the filaments were subjected to uniaxial drawing at moderate to high rates, depending upon the take-up speed on the winder. This uniaxial drawing introduced a strong axial flow creating molecular orientation along the drawing direction, leading to an increase in crystallization. This crystallization also changed the morphology

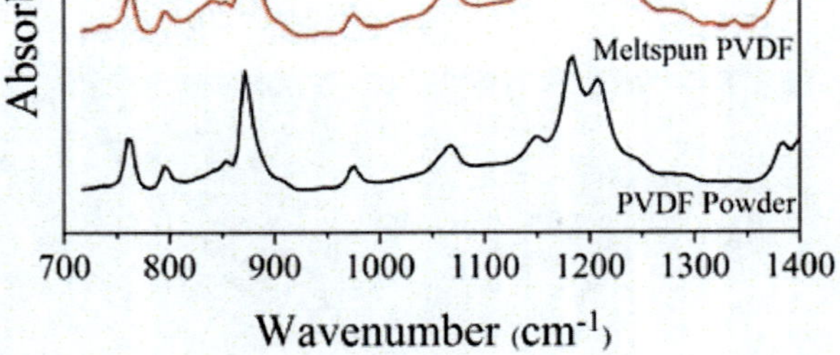

Fig. 3.5 FTIR results for PVDF powder, as-prepared meltspun and stretched/poled PVDF filament

Table 3.2 Percentage of β phase formation in all PVDF samples

Sample	A_α	A_β	F_β (%)
PVDF powder	0.3538	0.3324	42
Meltspun PVDF	0.14348	0.07319	28
Stretched and poled PVDF	0.2365	0.4572	60

of the structure from spherulitic (predominantly α phase and at low drawing ratios) to more fibril like (predominantly β phase and at high drawing ratios) at higher speeds [11]. To determine β phase crystallization in every sample, absorbance peaks related to the α and β phase at wavenumbers 764 and 840 cm^{-1} were evaluated. The level of β phase crystallization was calculated from Eq. (3.1). Here, A_α and A_β were absorbance for the α and β phase, respectively. X_α and X_β are the degree of crystallization for the α and β phases [12]. As can be seen from Table 3.2, the stretching and poling processes lead to an increase in β phase formation to 60% in the treated fibers compared to only 42% in the native PVDF powder. However, because of the melting process parameters and the collecting speed used, the β phase formation is lower in the melt-spun fibers at only 28% in comparison with the PVDF powder.

$$F_\beta = \frac{X_\beta}{X_\alpha + X_\beta} \times 100 = \frac{A_\beta}{A_\beta + 1.26A_\alpha} \times 100 \quad (3.1)$$

FTIR is a well-known method for the analysis and quantitative determination of the crystal structure in PVDF. According to the literature, the vibrational bands at 760, 796, 855, and 976 cm^{-1} are characteristic bands of the α phase, whereas the vibrational band at around 840 cm^{-1} denotes the β phase in PVDF [13].

As shown in the FTIR spectrum for the melt-spun PVDF filament in Fig. 3.5, the absorption bands corresponding to the α phase clearly appear in the spectrum and are quite strong. In contrast, the absorption bands which are attributed to the β phase are weak, which indicates that the α phase is dominant in the undrawn PVDF filaments. Results also showed that the degree of crystallinity in the filaments was determined by the melt draw ratio. Before the cold drawing process (at 80 °C) the crystalline structure of the fibers was predominantly in the α form after which the stretching and poling process lead to an increase in the β phase formation to 60%. The FTIR results indicated that drawing and poling processes could enhance β phase formation in the as-prepared stretched and poled PVDF fiber by more than 75% compared to PVDF powder.

To enhance the piezoelectric properties in the PVDF filament, the β-phase crystallinity should be present in higher concentrations since it displays the most effective piezoelectricity. Hence, X-ray diffraction (XRD) of the samples (meltspun filament, stretched and poled PVDF and the starting powder) was performed to evaluate the formation of this phase. Figure 3.6 illustrates the XRD diffractograms of a typical β crystal phase peak at around $2\theta = 20°$ which is assigned to the total diffraction in (110) and (200) planes. The peak at 18.30° corresponds to the reflection of the (020) plane of the α-phase [10]. The ratio intensity of β to α phase is increased from 1.21 to 1.58 for the meltspun PVDF compared to stretched and poled PVDF filaments. Hence, this may suggest that most of the crystal phase in stretched and poled PVDF filament has been changed to β-phase by the cold drawing (at the temperature of 80 °C) and poling process, whereby the β-phase crystal structure becomes dominant with some transitions or mixtures of α and β phases.

The ratio intensity for PVDF powder (1.63) is higher than meltspun filament (1.21) because the α phase peak intensity for the meltspun PVDF filament is higher than in

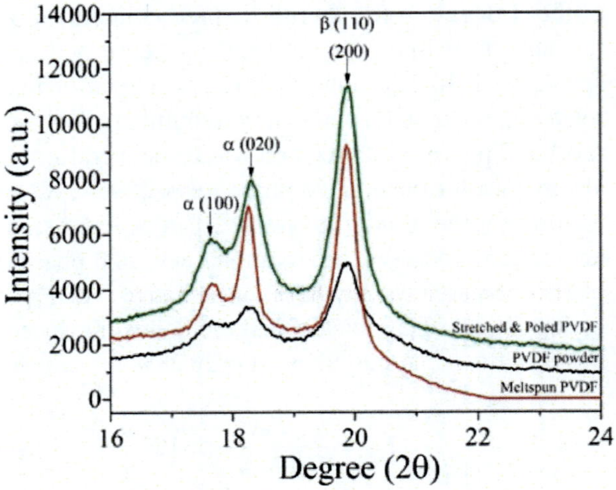

Fig. 3.6 X-ray diffraction patterns for the powder, as-prepared melt-spun, stretched and poled PVDF filament

the PVDF powder. As previously explained, during melt-spinning of the filaments the collecting speed was low which leads to higher α phase formation in the melt-spun fiber when compared to the powder. The level of β phase formation is related to the take-up speed of the collector and axial drawing of filaments and so this phase is more predominant in stretched and poled PVDF filaments. The X-ray diffraction patterns show that the intensity ratio of the β to α phase was increased from 1.21 to 1.58 in the as-prepared stretched and poled PVDF fiber compared to the as-spun PVDF fiber. In addition, thermal analysis of the PVDF fiber indicated that the formation of the β phase crystalline structure in the as-prepared PVDF fibers was found to be 67.5% and 71.2% for the as-spun PVDF fiber and stretched and poled PVDF fiber, respectively. Further investigation of the PVDF powder and filaments was carried out by measuring the melting temperature and melting enthalpy using DSC (Fig. 3.7). Based on the DSC thermograms, the melting temperature (T_m) of stretched and poled PVDF filaments (171 °C) was lower than the meltspun PVDF filaments (173 °C). This indicates that more β-phase crystals had been formed during the cold drawing process. The crystalline structure of β-phase PVDF has a lower melting temperature than α-phase PVDF [14]. It can also be seen from Fig. 3.8 that the melting enthalpy of stretched and poled PVDF filaments ($\Delta H_m = 74.46$ J g^{-1}) is more than the meltspun PVDF filaments (70.60 J g^{-1}). These results confirmed that the formation of the β phase crystalline structure in the stretched and poled PVDF filaments is higher than in the meltspun filaments (around 67.5% for meltspun PVDF filaments and 71.2% for stretched and poled PVDF filaments). Gheibi et al. [15] have done the same analysis in a comparison of β-phase formation in electrospun PVDF nanofiber mats and granules. The melting enthalpy of PVDF nanofibers ($\Delta H_m = 68.22$ J g^{-1}) is much more than that of the granules (41.70 J g^{-1}). As a result, there is an increase in crystallinity for electrospun nanofiber mats, which confirmed the formation of a new crystalline structure (β phase) in electrospun nanofiber mats [15]. The degree of crystallinity, ΔH_c was measured as the ratio between ΔH_m and ΔH_{100}, where ΔH_{100}

is the melting enthalpy of totally crystalline PVDF material ($\Delta H_{100} = 104.5$ J g^{-1} for α phase and $\Delta H_{100} = 103.4$ J g^{-1} for β-phase by using Eq. (3.2) as below [16]:

$$\Delta Hc = \frac{\Delta H_m}{\Delta H_{100}} \times 100 \tag{3.2}$$

In fiber production, melt spinning is generally followed by cold drawing (drawing at a temperature below the crystal melting point) of the fiber, as undrawn fibers exhibit poor mechanical properties, such as low stiffness, high strain to break, and high irreversible extensibility. Drawing is an irreversible elongation of an as-spun material in the solid state. Both cold-drawn and non-stretched PVDF filaments were mechanically tested to determine the effect of longitudinal stretching on the polymer filament (Fig. 3.8). Tenacity test results show that tensile strength of stretched PVDF filament is higher than non-stretched PVDF filament. Hadimani et al. [17] did mechanical test for poled and unpoled PVDF filament the results showed that higher force needed to break the poled PVDF fibers which proved that poling process increased the strength of PVDF fibers. Gomes et al. [18] have studied the effect of stretching ratio and temperature on α to β phase transformation in PVDF. They observe that variations in the phase content are accompanied by changes in the degree of crystallinity and the microstructure of the material, and these changes have a significant impact on the macroscopic piezoelectric and ferroelectric response of the material.

The mechanical properties and morphology of the as-prepared melt-spun PVDF fiber have been investigated. The mechanical properties of the as-prepared stretched PVDF fibers (Fig. 3.8) show that the ultimate tensile strength, elastic modulus and elongation at break were 110 MPa, 843 MPa and 58%, respectively, and they are significantly stiffer and stronger than the unstretched filament (Table 3.3). While the tensile strength and the elastic modulus of the as-spun PVDF fiber are similar to those of a previously reported PVDF fiber, the elongation at break of the as-prepared

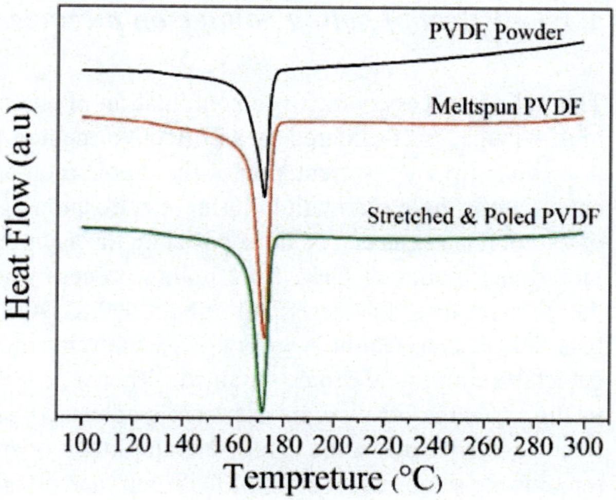

Fig. 3.7 DSC patterns for powder, melt-spun, stretched and poled PVDF filaments in triaxial braid structure

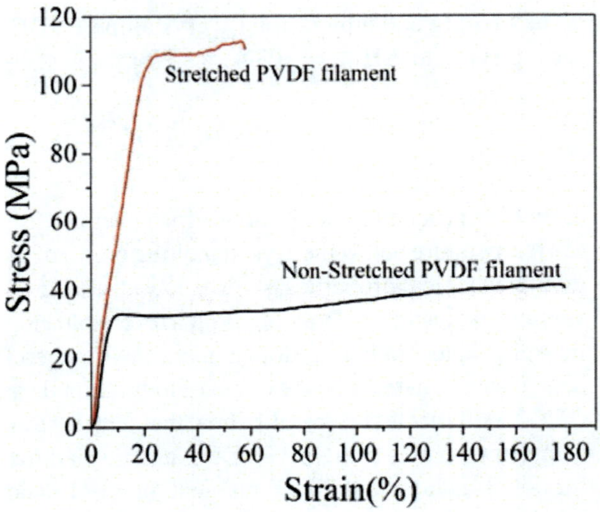

Fig. 3.8 Stress strain curves for the as-prepared stretched and non-stretched PVDF filament

Table 3.3 Measurement results taken from the Shimadzu tensile tester

	Stress (MPa)	Strain (%)	Young's modulus (MPa)
Non-stretched filament	37	178	354
Stretched filament	110	58	843

PVDF fiber was almost double that of the melt spun PVDF fiber reported by Leal et al. [19].

3.3.3 Effect of poling voltage on piezoelectric response

The poling process, which is a combination of extension, heating, and exposure to high voltage, is considered as an effective method to improve the power generation capability by re-orientation of the dipole moments. Mandal et al. [20] used IR analysis of dipole orientation during electrospinning. The results show that dipole orientation analysis can be done based on the asymmetric and symmetric stretching vibrational modes of CH_2. Here poling process was done for PVDF fibers along the filament length as described in Sect. 2.2.8. During the PFM test, when the DC bias field is zero (Point A in Fig. 3.9), the remnant polarization is positive, and a detectable amplitude can be obtained. When a positive DC field is applied (Point B in Fig. 3.9), the positive polarization is enhanced, since the PVDF electrets experience a barrier discharge. Thus, the amplitude is greatly increased, and the phase remains the same as that at Point A. When the voltage decreases, the amplitude will also decrease due to an inverse barrier discharge in the PVDF filament. When the

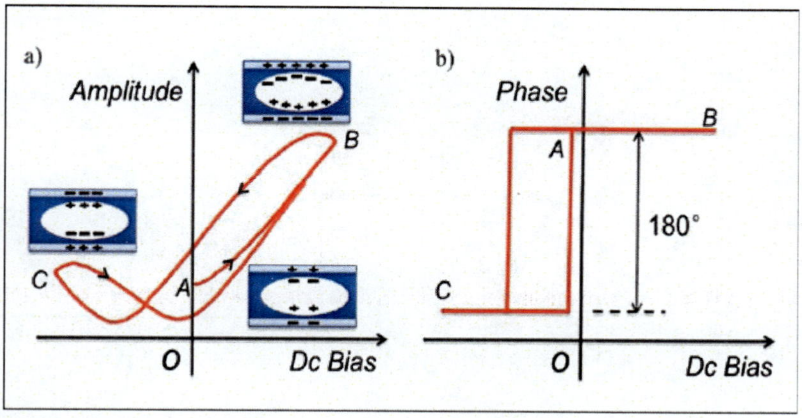

Fig. 3.9 Mechanism of amplitude butterfly curve and phase hysteresis Loop of cellular PP thin film during SS-PFM testing [21]

DC field changes to negative (Point C in Fig. 3.9), the polarization reverses to be negative but its magnitude is considerably smaller than the value at Point B as a result of the large positive internal bias field induced during corona poling. Therefore, the amplitude at Point C is considerably smaller than that at Point B, leading to the strongly asymmetric amplitude butterfly curve shown in Fig. 3.9a. Meanwhile, as the polarization direction at Point C is opposite to that at Point B, the phase difference [21] is 180°, which is also close to the experimental observations in Fig. 3.10. To evaluate the extent of penetration of the poling voltage along the fiber length and its effectiveness, PFM testing was carried out using the 10 mm PVDF fiber poled with 9 kV. PFM amplitude responses as a function of distance along the fiber from the positive poling electrode are shown in Fig. 3.11a. As can be seen from Fig. 3.11a, the piezoelectric response was sensitive to the distance from the electrode such that the maximum output amplitude for the poled PVDF filament occurred at less than 1 mm distance from the positive poling electrode. Consequently, PFM characterization of the poled PVDF fibers of more than 40 mm in length was not carried out due to insufficient piezo responses. The results of PFM phase responses as a function of fiber length for the sample with 10 mm length in the vertical direction are shown in Fig. 3.11b. The amplitude versus Vdc loop was measured by applying a voltage of −25 to +25 V to the tip with respect to the ground. As can be seen from Fig. 3.11b, the amplitude versus *V*dc curves are hysteretic and the shape of the loops strongly resembles the butterfly loop normally observed in piezoelectrics [7]. However the results of piezoresponse in the horizontal direction were insignificant. The phase of the PFM response signal is directly related to the direction of the electric polarization of the microscopic region of the PVDF fiber under the tip. Thus, the hysteretic switching of the response signal phase by 180° in response to a sweeping DC voltage is attributed to the switching of the direction of polarization of the PVDF dipoles along the direction of the electric field nearly symmetrical butterfly-shaped amplitude curves (Fig. 3.11). However, the PFM results for the poled PVDF fiber with

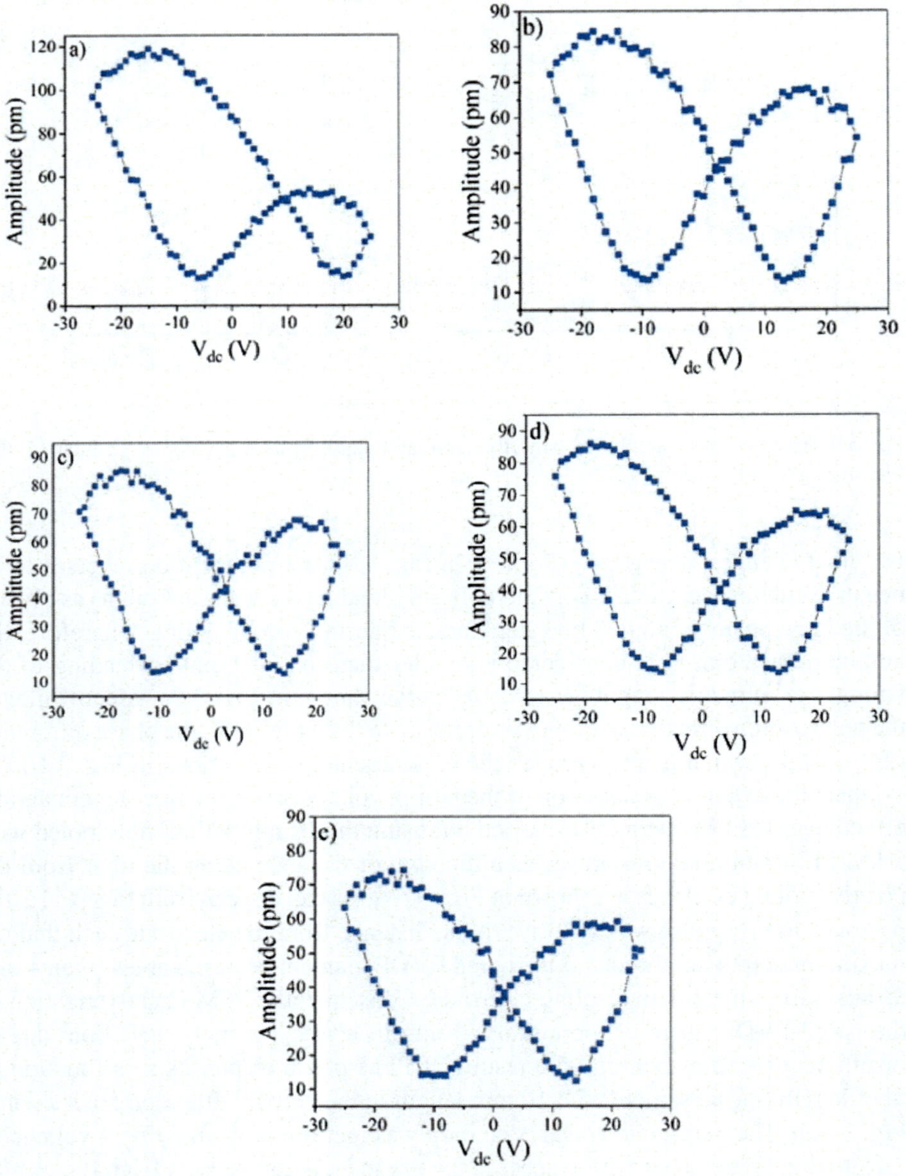

Fig. 3.10 Amplitude butterfly curve test results for the as-prepared poled PVDF fiber with 10 mm length at 5 positions along the length from **a** to **e** respectively

20 mm length in Fig. 3.11c shows an asymmetric butterfly amplitude curve. These results confirm that the poling process was most effective when the fiber length was less than 10 mm. Microstructural images of poled and unpoled PVDF filaments were captured at the same magnifications and compared. Figure 3.12 shows SEM images of (a) the unpoled and (b) the poled PVDF filaments. Although some authors have

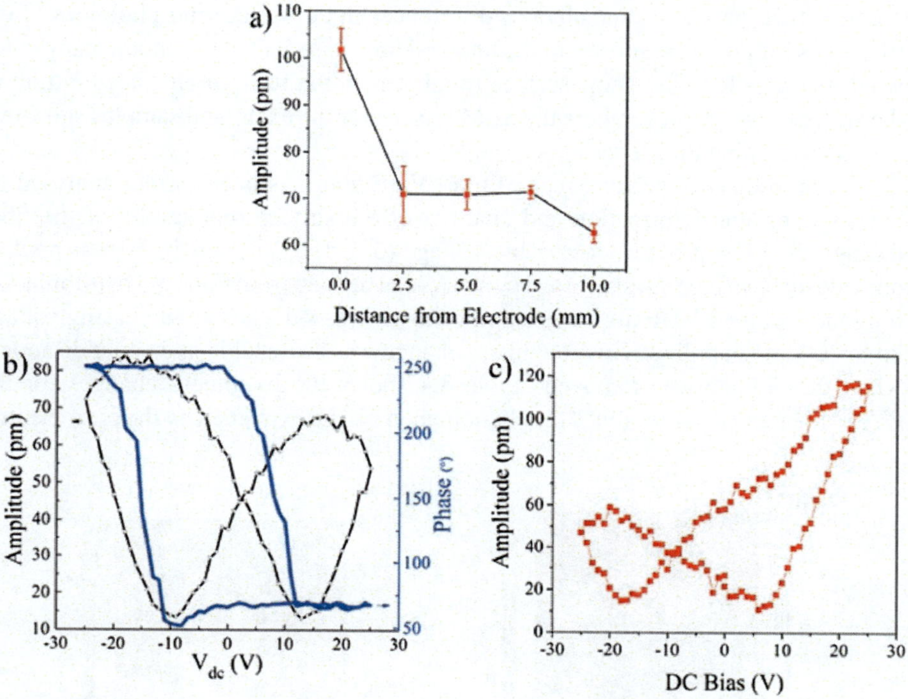

Fig. 3.11 PFM amplitude responses as a function of fiber length for **a** the as-prepared 10 mm poled PVDF fiber, **b** local PFM phase hysteresis loop and piezoelectric butterfly loop of the as-prepared poled PVDF fiber with 10 mm length and 200 μm thickness by switching the voltage from −25 to +25 V (±1.5 kV cm^{-1} electric field) and **c** asymmetric amplitude butterfly curve for 20 mm length

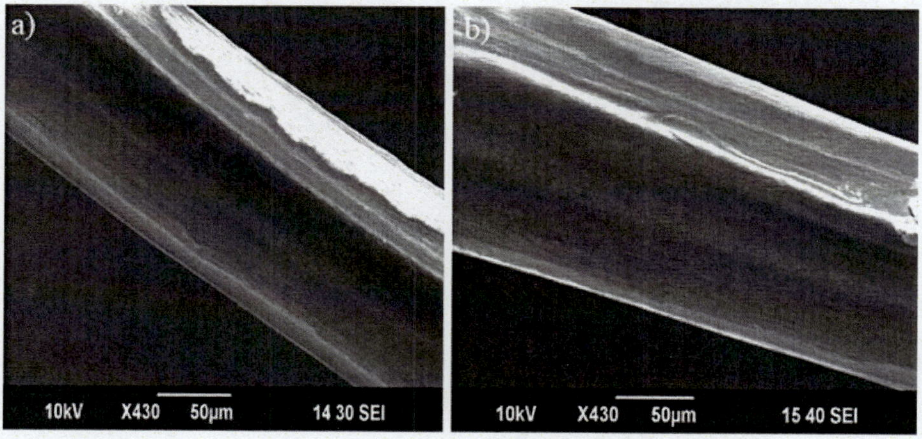

Fig. 3.12 SEM images of PVDF filament surfaces: **a** unpoled, **b** poled sample

reported that there are some obvious differences in the surface morphologies of the two fibres [22] with the surface of the unpoled fibre appears not to contain many voids whereas the poled fibre show surface roughness in the form of enhanced pitting of the surface [22]. However, here the SEM images for the examined samples show that there is no visible difference.

The piezoelectric performance of the PVDF fiber has been further analysed by investigating the deformation and changes of the surface topography of the fiber through the PFM process. A variable voltage (0, 2.5, 5, 7.5, and 10 V) was applied onto the fiber surface at a single test point. As can be seen from Fig. 3.13 and Table 3.4, roughness of the PVDF fibre was significantly increased when an increasing voltage up to 10 V was applied. This roughness change was switchable and could disappear when the voltage was removed. Table 3.4 shows the roughness changes for the PVDF fiber and its vibration. This phenomenon can be explained by the piezoelectric

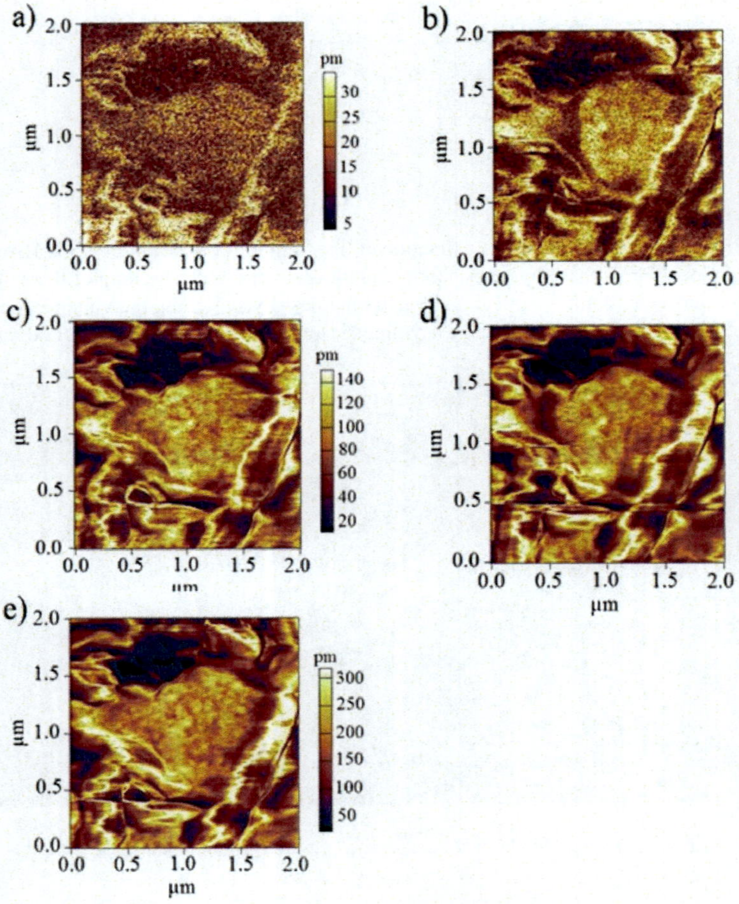

Fig. 3.13 PFM topography images of the as-prepared PVDF fiber as a function of applied voltage: **a** 0, **b** 2.5, **c** 5, **d** 7.5 and **e** 10 V

Table 3.4 Deformation (surface topography) of the as-prepared PVDF filament as a function of applied voltage

Applied voltage (V)	Deformation (pm)
0	7.642
2.5	14.481
5	34.298
7.5	54.074
10	72.372

behaviour of the PVDF fiber and confirmed piezoelectric response of the as-prepared PVDF fiber.

3.3.4 *Fabrication of Wearable Energy Harvesters*

A new strategy of a triaxial architecture based on a piezoelectric PVDF fiber, silver coated nylon and braiding technology has been developed to manufacture continuous lengths of the wearable energy harvesting generator. The triaxial braided PVDF structure fabricated here has a number of practical advantages including the ability for mass production, a practical architecture which is suitable for poling and charge collection for unlimited length energy harvesting devices while also being flexible and robust with the outer braid providing more protection from general physical damage.

As the PFM results indicate that a high performance piezoelectric PVDF fiber requires short distances between poling electrodes, the design adopted here promotes poling along the fiber radial direction. The developed triaxial braided structure enables poling between the inner and outer electrodes where the PVDF fiber is as an intermediate structure (Fig. 3.14). The direction of the applied force and the direction of the produced dipoles are the same when the structure is compressed, hence making d_{33} the active mode (transverse mode) [23]. The triaxial braided PVDF fiber has been successfully fabricated using the melt-extruded PVDF fiber and commercial silver coated nylon. These braided structures were exceptionally flexible and lengths of up to 100 m were prepared and only limited by the availability of the PVDF fiber.

3.3.5 *Mechanical Energy Harvesting Performance*

To characterize the mechanical energy harvesting performances of the developed textile piezoelectric generators an impact test was carried out by dropping a ball from two different heights onto the surface of the braided devices that was fixed to a rigid structure. The dropped ball bounced a number of times before coming to rest and each impact generated a voltage peak with the peak intensity diminishing with time. A 2 cm long braided energy harvesting sample produced an open-circuit voltage of

Fig. 3.14 Photograph of the cross-section of the fabricated energy harvesting generator based on triaxial braided PVDF fibers; **a**, **b** silver coated nylon as outer and inner electrodes, respectively, and **c** braided piezoelectric PVDF fiber

~230 mV (Fig. 3.15a) and ~380 mV (Fig. 3.15b) at an applied pressure of 0.017 MPa and 0.023 MPa, respectively. The pressure was the peak pressure measured using a pressure sensor (Flexiforce Pressure Sensor—25 lbs) located underneath the sample. As can be seen from Fig. 3.15a, b, from each drop event onto a sample surface, two peaks appear in the output voltage: the first corresponds to the applied force, resulting in a negative voltage output, while the second corresponds to the relaxation of the sample after the load has been removed, resulting in a positive voltage output. The intensity of the generated voltage peaks decreased over time as the applied energy from the bouncing weight dissipated as it came to rest. In addition, the developed triaxial braided PVDF structure showed asymmetric output voltages and provides more negative voltage during the impact test. This phenomenon could be explained by synergistic triaxial-architecture design which enhances the contact area between the PVDF braid and electrodes during the impact test that would be capable of transferring more negative charge after compressing. The voltage output of the developed triaxial braided energy generator as a function of applied impact pressure is shown in Fig. 3.15c. The voltage output increases up to 465 mV with increasing the applied impact pressure up to 0.09 MPa. It was observed that beyond a certain threshold of impact (0.09 MPa), the voltage output from the triaxial braided energy harvester was abruptly reduced to zero. In fact, the lower resilience of the triaxial braided structure was momentarily compressed and sheared to such an extent that the two opposite conducting ends shorted out each other and hence no output was observed. The power output of the developed triaxial energy harvesting textiles has been measured using an impact pressure of 0.023 MPa, which provided a reproducible response without

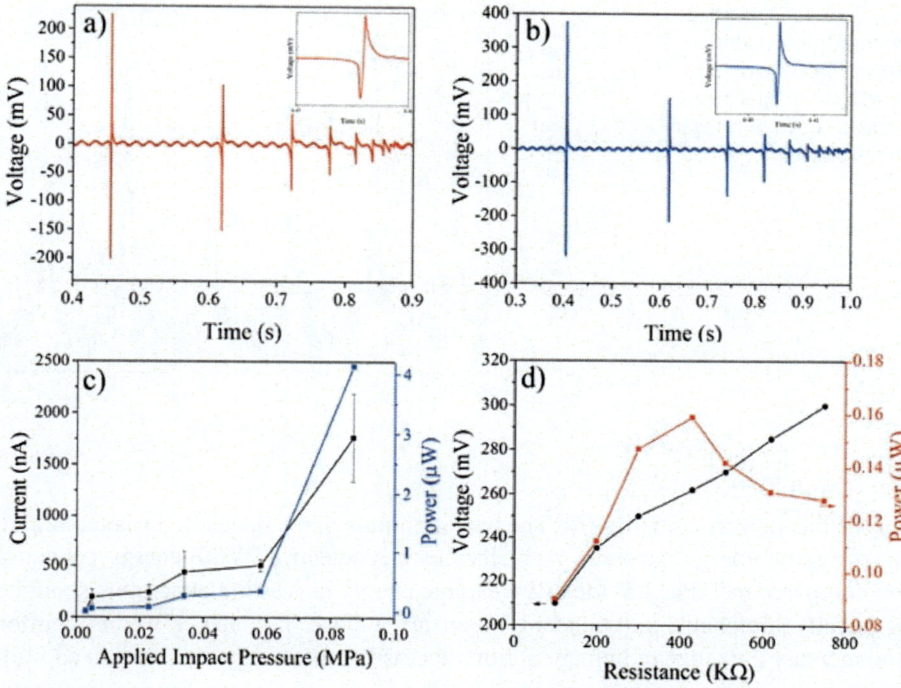

Fig. 3.15 The mechanical energy harvesting performance of developed triaxial braided piezoelectric fibers: **a**, **b** output voltage for the 2 cm fabricated device at an applied impact pressure of 0.017 MPa and 0.023 MPa, respectively. The inset shows voltage output time for the first peak. **c** Variation of current and peak power output as a function of applied impact pressure for the developed triaxial piezo energy generator and **d** the peak values of the power and voltage for the fabricated device (obtained at an impact pressure of 0.023 MPa)

causing any damage. Regards to the relationship between load resistance (R) and output (V) ($P = V^2/R$) by increasing load the voltage will increase but after 400 KΩ, load increasing rate is more than voltage with leads to lower power output. As can be seen from Fig. 3.15d, the piezoelectric textile generator provided a maximum power output of 0.16 μW which is corresponding to a voltage output of 380 mV at an applied impact pressure of 0.023 MPa. This applied impact pressure is ~22% lower than previously reported for the 3-D spacer piezoelectric textiles [11]. An in-house setup was constructed to apply a continuous pressure of 0.0145 MPa on the sample and a bridge rectifier (4 diodes of 1N4004) was placed in the circuit to feed a capacitor. The generated voltage after rectification is 350 mV (Fig. 3.16) and its power is 52.59 μW cm^{-3}.

The power output from the developed triaxial braided PVDF energy generator is much higher than those reported for energy generators based on PVDF films and fibers. The power density of the developed piezoelectric textile energy generator was found to be 29.62 μW cm^{-3} (5700 μW kg^{-1}) based on the volume and mass of the PVDF fibers. The value is ~1559% higher than previously reported (1.78 μW cm^{-3})

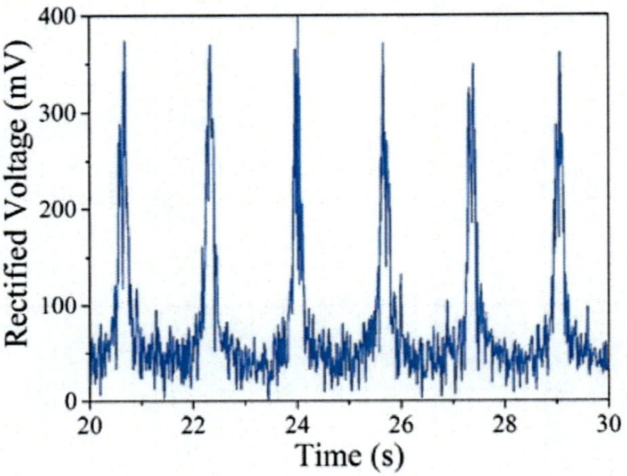

Fig. 3.16 The rectified voltage from continues impact pressure of the developed triaxial piezo braided structure using an in-house setup

for the piezoelectric textiles [5]. The power outputs of the developed triaxial braided piezoelectric energy harvester and pervious piezoelectric PVDF energy generators are compared in Table 3.5. Ideally, wearable energy harvesting generators should be elastically stretchable and bendable to ensure a close fit, enhance wearer comfort, and increase the range of human motions accessible for energy recovery. In addition, the stability of the energy harvesting textiles is a critical issue for practical applications. Consequently, we performed a bending cycling test with a triaxial braided energy harvesting generator to evaluate the effect of bending on the voltage output and stability of the developed device. The bending cycling test was carried out using a periodic compression strain test where the end-to-end distance was reduced by buckling to 50% at a deformation rate of 300 mm min^{-1} over 1000 cycles. As can be seen from Fig. 3.17a, the output voltage was found to be 150 mV with a power density of 4.62 μW cm^{-3} or 1090 μW kg^{-1} for up to 1000 bending cycles at a frequently of 0.6 Hz. These results indicated that the triaxial textile energy generator is extraordinarily flexible and stable for thousands of cycles with no change in its performance (Fig. 3.17b). The bending deformation of the energy generator provided a lower voltage and power output (~150 mV and 4.62 μW cm^{-3}) compared with the impact test (~380 mV and 29.62 μW cm^{-3}) due to the greater stress exerted on the sample fibres during the impact test. It is recognised that the low energy density of piezoelectric devices at low frequency and static load sources have limited their application. The lowest frequency reported for PVDF devices has been limited to between 1 and 10 Hz [45]. Our results indicated that the developed triaxial braided energy generator overcomes this limitation with significant energy harvesting cycling at very low speed (0.25 Hz). This frequency is comparable to that of normal human walking, so that our textile energy harvester could be utilized for diverse applications such as self-powered wireless sensors and powering wearable electronic devices (i.e. electronic textiles). The sensitivity of the as-prepared triaxial piezoelectric textiles was evaluated using output voltage divided by the applied force [46]. The results indicated that the triaxial piezo energy generator exhibited an almost 4-fold increase

Table 3.5 Comparison of the power output of the energy generator based on piezoelectric PVDF films and fibers

Material	Type	Power	Power density	References
ZnO + CNT	Film casting	18.75 μW	–	[24]
Alkaline niobate-based particles (KNLN)	Film	0.5 mW	–	[25]
PVDF-HFP-TEA-BF4	Yarn	–	43 μW h/cm^2	[26]
PVDF	Filament yarn	0.1 mW	–	[5]
PVDF	3D spacer fabric	–	1.10–5.10 μW/cm^2	[11]
P(VDF-TrFE)	Electrospun webs	–	5.9 mW/cm^3	[27]
PVDF	Monofilament	2.79 nW	–	[28]
PVDF	Nanofiber	–	–	[29]
PVDF	Film	–	65 nW/cm^2	[30]
PVDF–$NaNbO_3$	Nanofilm	14.96 μW	16.2 μW/cm^3	[23]
$BaTiO_3$/P(VDF-TrFE)	Electrospun membrane	0.02–25 μW	–	[31]
$BaTiO_3$	Thin film	–	7 mW/cm^3	[32]
P(VDF-TrFE) and $BaTiO_3$	Electrospun fiber	–	16 μW/cm^2	[33]
PVDF	Meltspun	0.7 mW	–	[34]
PVDF + CB	Meltspun	15 nW	–	[35]
PVDF + ZnO	Film (hybrid device in plane-shape)	–	16 μW/cm^3	[36]
PVDF	Nanofiber	0.12 nW	–	[37]
PVDF + ZnO	Film and nanowires	–	0.17 mW/cm^3	[38]
PVDF + PZT	Film	1.3 mW	–	[39]
PVDF + BTZO	Film	15.8 nW	2.52 nW/cm^2	[40]
PVDF-niobate-based	Film	–	11.7 μW/cm^2	[41]
PVDF + activated carbon	Film	–	63.07 mW/m^2	[42]
PVDF + graphitic carbon nitrid	Film	–	100 mW/cm^2	[43]
PVDF + GO + AlO	Nano composite film		27.97 μW/cm^3	[44]
PVDF	Triaxial braided piezo fibers	–	29.62 μW/cm^3	This work

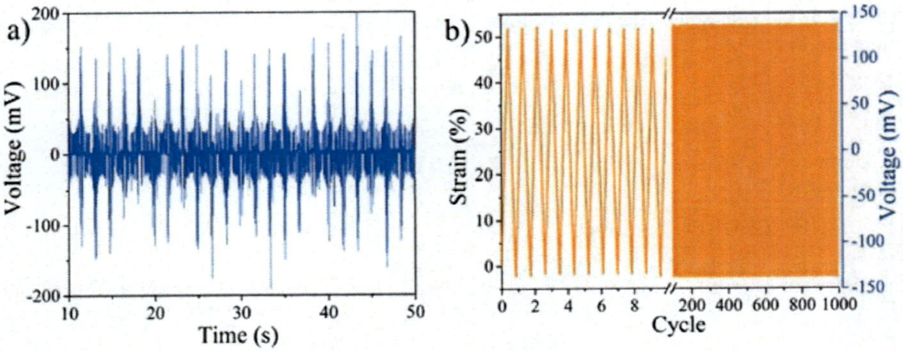

Fig. 3.17 Characteristics of the developed triaxial braided piezoelectric textile during the bending cycling test: **a** the voltage output of the triaxial textile energy generator with a maximum strain of 50% and at 0.25 Hz and **b** the stability of performance of the triaxial braided piezoelectric textile in the bending test during 1000 cycles to a maximum strain of 50% and at 0.6 Hz

in the value (1.6 V/N) compared to those previously reported [25]. Furthermore, we have explored the scaling of the triaxial textile where multiple mechanical energy harvesters are assembled to meet specific energy and power needs for practical applications. As shown in Fig. 3.18, a fabricated woven textile energy generator based on triaxial braided piezo fibers with five devices in parallel was assembled. This device provided a voltage output of ~380 mV and a power output of 0.8 μW at an applied pressure of 0.023 MPa. The textile energy generator can be integrated with a soft and stretchable silicone material to provide protection from the environment including washing. Interestingly, the voltage output of the silicone coated triaxial braided energy generator was enhanced by more than 108% compared to that of the device without coating. This improvement may be due to better packaging of fibers within the silicone coated device which could improve the efficiency of the charge collection as well as supporting the mechanical performance of the device. Also as a demonstration, the developed triaxial braided textile energy generator can charge a commercial battery or capacitor from tensile and bending deformations.

3.4 Conclusion

In summary, a new strategy of synergistic triaxial-architecture design based on piezoelectric fibers, conducting fibers, and braiding technology is developed to manufacture a new class of wearable energy harvesting generators. A novel triaxial piezoelectric textile energy harvester has been developed with a view to enhance both the piezoelectric performance and mechanical properties of the developed energy harvesting generator through a fabrication process. The triaxial braided energy generator has been fabricated using high-performance PVDF fibers as the piezoelectric polymer and silver coated nylon as inner and outer electrodes though a melt-spinning and

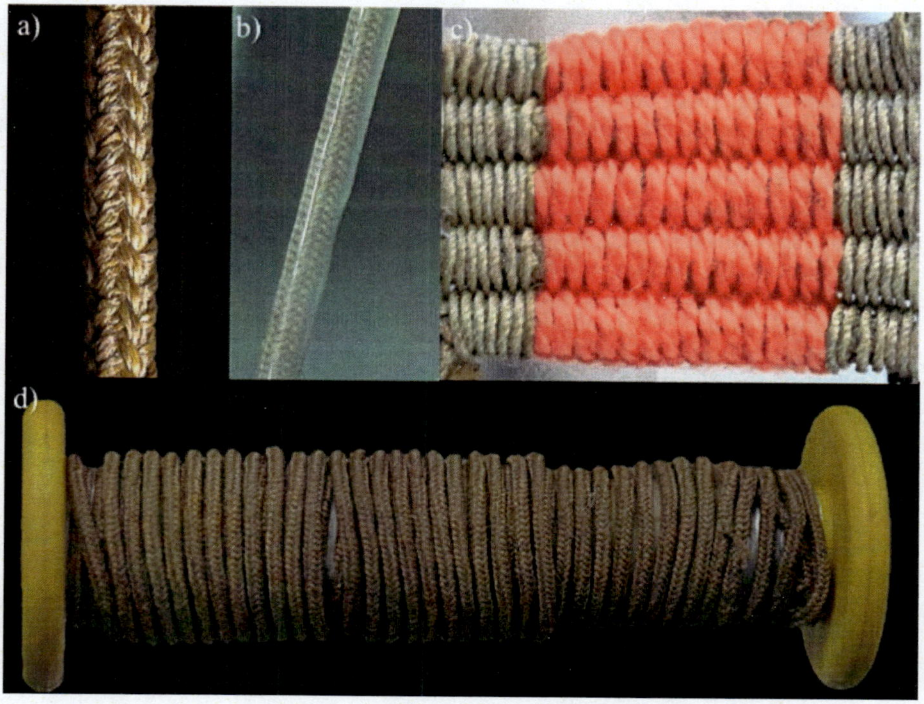

Fig. 3.18 Photographs of **a** the fabricated triaxial braided energy harvester based on meltspun PVDF and silver coated fibers, **b** silicone coated triaxial braided energy generator, **c** fabricated woven textile energy generator based on triaxial braided piezo fibers with five devices in parallel, and **d** a spool of the as-prepared triaxial braided piezoelectric energy harvesters

braiding process. We demonstrate novel triaxial braided textile harvesters that piezoelectrically convert tensile mechanical energy into electrical energy. Compressing or bending braided PVDF yarns generated a maximum output voltage of 380 mV and a power density of 29.62 μW cm^{-3} which is ~1559% higher than previously reported (1.78 μW cm^{-3}) for piezoelectric textiles. It is found that the developed triaxial energy generator exhibits significantly higher sensitivity by a factor of 4 in compared with the PVDF energy generator. Unlike other piezoelectric harvesters, the triaxial braided PVDF yarn achieves tensile energy harvesting and shows extreme durability which enables cycling with up to 50% strain for thousands of cycles with no changes in its performance. Finally, the fabricated triaxial braided piezoelectric energy harvesting textile provided an efficient and novel way to overcome the stability issues due to the poor fatigue resistance of the metallic electrodes. The triaxial braided piezofiber is strong, lightweight and exceptionally flexible and it is expected to be applicable to wearable devices where high performance is necessary.

References

1. Donelan, J.M., Li, Q., Naing, V., Hoffer, J.A., Weber, D.J., Kuo, A.D.: Biomechanical energy harvesting: generating electricity during walking with minimal user effort. Science **319**(5864), 807–810 (2008)
2. Lee, S., Bae, S.-H., Lin, L., Yang, Y., Park, C., Kim, S.-W., Cha, S.N., Kim, H., Park, Y.J., Wang, Z.L.: Super-flexible nanogenerator for energy harvesting from gentle wind and as an active deformation sensor. Adv. Funct. Mater. **23**(19), 2445–2449 (2013)
3. Parangusan, H., Ponnamma, D., Al-Maadeed, M.A.A.: Stretchable electrospun PVDF-HFP/Co-ZnO nanofibers as piezoelectric nanogenerators. Sci. Rep. **8**(1), 754 (2018)
4. Garain, S., Sinha, T.K., Adhikary, P., Henkel, K., Sen, S., Ram, S., Sinha, C., Schmeißer, D., Mandal, D.: Self-poled transparent and flexible UV light-emitting cerium complex–PVDF composite: a high-performance nanogenerator. ACS Appl. Mater. Interfaces **7**(2), 1298–1307 (2015)
5. Lund, A., Rundqvist, K., Nilsson, E., Yu, L., Hagström, B., Müller, C.: Energy harvesting textiles for a rainy day: woven piezoelectrics based on melt-spun PVDF microfibres with a conducting core. NPJ Flex. Electron. **2**(1), 9 (2018)
6. Guan, Z., Jiang, Z.-Z., Tian, B.-B., Zhu, Y.-P., Xiang, P.-H., Zhong, N., Duan, C.-G., Chu, J.-H.: Identifying intrinsic ferroelectricity of thin film with piezoresponse force microscopy. AIP Adv. **7**(9), 095116 (2017)
7. Maji, S., Sarkar, P.K., Aggarwal, L., Ghosh, S.K., Mandal, D., Sheet, G., Acharya, S.: Self-oriented β-crystalline phase in the polyvinylidene fluoride ferroelectric and piezo-sensitive ultrathin Langmuir-Schaefer film. Phys. Chem. Chem. Phys. **17**(12), 8159–8165 (2015)
8. Liu, X., Ma, J., Wu, X., Lin, L., Wang, X.: Polymeric nanofibers with ultrahigh piezoelectricity via self-orientation of nanocrystals. ACS Nano **11**(2), 1901–1910 (2017)
9. Rome, L.C., Flynn, L., Goldman, E.M., Yoo, T.D.: Generating electricity while walking with loads. Science **309**(5741), 1725–1728 (2005)
10. Lund, A., Hagström, B.: Melt spinning of β-phase poly(vinylidene fluoride) yarns with and without a conductive core. J. Appl. Polym. Sci. **120**(2), 1080 (2011)
11. Soin, N., Shah, T.H., Anand, S.C., Geng, J., Pornwannachai, W., Mandal, P., Reid, D., Sharma, S., Hadimani, R.L., Bayramol, D.V.: Novel "3-D spacer" all fibre piezoelectric textiles for energy harvesting applications. Energy Environ. Sci. **7**(5), 1670–1679 (2014)
12. Mokhtari, F., Shamshirsaz, M., Latifi, M., Asadi, S.: Comparative evaluation of piezoelectric response of electrospun PVDF (polyvinilydine fluoride) nanofiber with various additives for energy scavenging application. J. Text. Inst. **108**(6), 906 (2017)
13. Mokhtari, F., Latifi, M., Shamshirsaz, M.: Electrospinning/electrospray of polyvinylidene fluoride (PVDF): piezoelectric nanofibers. J. Text. Inst. **107**(8), 1037 (2016)
14. Mokhtari, F., Shamshirsaz, M., Latifi, M.: Investigation of β phase formation in piezoelectric response of electrospun polyvinylidene fluoride nanofibers: LiCl additive and increasing fibers tension. Polym. Eng. Sci. **56**(1), 61 (2016)
15. Gheibi, A., Latifi, M., Merati, A.A., Bagherzadeh, R.: Piezoelectric electrospun nanofibrous materials for self-powering wearable electronic textiles applications. J. Polym. Res. **21**(7), 1 (2014)
16. Costa, C.M., Rodrigues, L.C., Sencadas, V., Silva, M.M., Rocha, J.G., Lanceros-Méndez, S.: Effect of degree of porosity on the properties of poly(vinylidene fluoride–trifluorethylene) for Li-ion battery separators. J. Membr. Sci. **407–408**, 193 (2012)
17. Hadimani, R.L., Characterisation of energy generating polyvinylidene fluoride (PVDF) based piezoelectric filament. Adv. Mater. Res. 410 (2012)
18. Gomes, J., Nunes, J.S., Sencadas, V., Lanceros-Mendez, S.: Influence of the β-phase content and degree of crystallinity on the piezo- and ferroelectric properties of poly(vinylidene fluoride). Smart Mater. Struct. **19**(6), 065010 (2010)
19. Leal, A.A., Mohanty, G., Reifler, F.A., Michler, J., Hufenus, R.: Mechanical response of melt-spun amorphous filaments. Sci. Technol. Adv. Mater. **15**(3), 035016 (2014)

20. Mandal, D., Yoon, S., Kim, K.J.: Origin of piezoelectricity in an electrospun poly(vinylidene fluoride-trifluoroethylene) nanofiber web-based nanogenerator and nano-pressure sensor. Macromol. Rapid Commun. **32**(11), 831–837 (2011)
21. Miao, H., Sun, Y., Zhou, X., Li, Y., Li, F.: Piezoelectricity and ferroelectricity of cellular polypropylene electrets films characterized by piezoresponse force microscopy. J. Appl. Phys. **116**(6), 066820 (2014)
22. Hadimani, R.L., Bayramol, D.V., Sion, N., Shah, T., Qian, L., Shi, S., Siores, E.: Continuous production of piezoelectric PVDF fibre for e-textile applications. Smart Mater. Struct. **22**(7), 075017 (2013)
23. Zeng, W., Tao, X.-M., Chen, S., Shang, S., Chan, H.L.W., Choy, S.H.: Highly durable all-fiber nanogenerator for mechanical energy harvesting. Energy Environ. Sci. **6**(9), 2631–2638 (2013)
24. Sun, H., Tian, H., Yang, Y., Xie, D., Zhang, Y.-C., Liu, X., Ma, S., Zhao, H.-M., Ren, T.-L.: A novel flexible nanogenerator made of ZnO nanoparticles and multiwall carbon nanotube. Nanoscale **5**(13), 6117 (2013)
25. Kyu, J.C., Kwi-Il, P., Jungho, R., Geon-Tae, H., Jae, L.K.: Large-area and flexible lead-free nanocomposite generator using alkaline niobate particles and metal nanorod filler. Adv. Funct. Mater. **24**(18), 2620 (2014)
26. Choi, C., Park, J.W., Kim, K.J., Lee, D.W., de Andrade, M.J., Kim, S.H., Gambhir, S., Spinks, G.M., Baughman, R.H., Kim, S.J.: Weavable asymmetric carbon nanotube yarn supercapacitor for electronic textiles. RSC Adv. **8**(24), 13112–13120 (2018)
27. Lang, C., Fang, J., Shao, H., Wang, H., Yan, G., Ding, X., Lin, T.: High-output acoustoelectric power generators from poly(vinylidenefluoride-co-trifluoroethylene) electrospun nano-nonwovens. Nano Energy **35**, 146 (2017)
28. Matsouka, D., Vassiliadis, S., Prekas, K., Bayramol, D.V., Soin, N., Siores, E.: On the measurement of the electrical power produced by melt spun piezoelectric textile fibres. J. Electron. Mater. **45**(10), 5112 (2016)
29. Wang, Z., Tan, L., Pan, X., Liu, G., He, Y., Jin, W., Li, M., Hu, Y., Gu, H.: Self-powered viscosity and pressure sensing in microfluidic systems based on the piezoelectric energy harvesting of flowing droplets. ACS Appl. Mater. Interfaces **9**(34), 28586 (2017)
30. Vatansever, D., Hadimani, R., Shah, T., Siores, E.: An investigation of energy harvesting from renewable sources with PVDF and PZT. Smart Mater. Struct. **20**(5), 055019 (2011)
31. Nunes-Pereira, J., Sencadas, V., Correia, V., Rocha, J.G., Lanceros-Méndez, S.: Energy harvesting performance of piezoelectric electrospun polymer fibers and polymer/ceramic composites. Sens. Actuators A **196**, 55 (2013)
32. Park, K.-I., Xu, S., Liu, Y., Hwang, G.-T., Kang, S.-J.L., Wang, Z.L., Lee, K.J.: Piezoelectric $BaTiO_3$ thin film nanogenerator on plastic substrates. Nano Lett. **10**(12), 4939 (2010)
33. Dhakras, D., Ogale, S.: High-performance organic-inorganic hybrid piezo-nanogenerator via interface enhanced polarization effects for self-powered electronic systems. Adv. Mater. Interfaces **3**(20), 1600492 (2016)
34. Nilsson, E., Mateu, L., Spies, P., Hagström, B.: Energy harvesting from piezoelectric textile fibers. Proc. Eng. **87**, 1569 (2014)
35. Nilsson, E., Lund, A., Jonasson, C., Johansson, C., Hagström, B.: Poling and characterization of piezoelectric polymer fibers for use in textile sensors. Sens. Actuators A **201**, 477–486 (2013)
36. Lee, M., Chen, C.-Y., Wang, S., Cha, S.N., Park, Y.J., Kim, J.M., Chou, L.-J., Wang, Z.L.: A hybrid piezoelectric structure for wearable nanogenerators. Adv. Mater. **24**(13), 1759–1764 (2012)
37. Qi, Y., McAlpine, M.C.: Nanotechnology-enabled flexible and biocompatible energy harvesting. Energy Environ. Sci. **3**(9), 1275 (2010)
38. Cha, S., Kim, S.M., Kim, H., Ku, J., Sohn, J.I., Park, Y.J., Song, B.G., Jung, M.H., Lee, E.K., Choi, B.L.: Porous PVDF as effective sonic wave driven nanogenerators. Nano Lett. **11**(12), 5142 (2011)
39. Puscasu, O., Counsell, N., Herfatmanesh, M.R., Peace, R., Patsavellas, J., Day, R.: Powering lights with piezoelectric energy-harvesting floors. Energy Technol. **6**(5), 906 (2018)

40. Alluri, N.R., Saravanakumar, B., Kim, S.-J.: Flexible, hybrid piezoelectric film ($BaTi_{(1-x)}Zr_xO_3$)/PVDF nanogenerator as a self-powered fluid velocity sensor. ACS Appl. Mater. Interfaces **7**(18), 9831–9840 (2015)
41. Zhang, C., Fan, Y., Li, H., Li, Y., Zhang, L., Cao, S., Kuang, S., Zhao, Y., Chen, A., Zhu, G., Wang, Z.L.: Fully rollable lead-free poly(vinylidene fluoride)-niobate-based nanogenerator with ultra-flexible nano-network electrodes. ACS Nano **12**(5), 4803–4811 (2018)
42. Alluri, N.R., Chandrasekhar, A., Jeong, J.H., Kim, S.-J.: Enhanced electroactive β-phase of the sonication-process-derived PVDF-activated carbon composite film for efficient energy conversion and a battery-free acceleration sensor. J. Mater. Chem. C **5**(20), 4833–4844 (2017)
43. Kumar, S., Jena, A., Hu, Y.C., Liang, C., Zhou, W., Hung, T.F., Chang, W.S., Chang, H., Liu, R.S.: Cobalt diselenide nanorods grafted on graphitic carbon nitride: a synergistic catalyst for oxygen reactions in rechargeable $Li-O_2$ batteries. ChemElectroChem **5**(1), 29–35 (2018)
44. Karan, S.K., Bera, R., Paria, S., Das, A.K., Maiti, S., Maitra, A., Khatua, B.B.: An approach to design highly durable piezoelectric nanogenerator based on self-poled PVDF/AlO-rGO flexible nanocomposite with high power density and energy conversion efficiency. Adv. Energy Mater. **6**(20), 1601016 (2016)
45. Cannarella, J., Arnold, C.B.: Toward low-frequency mechanical energy harvesting using energy-dense piezoelectrochemical materials. Adv. Mater. **27**(45), 7440–7444 (2015)
46. Katsouras, I., Asadi, K., Li, M., van Driel, T.B., Kjær, K.S., Zhao, D., Lenz, T., Gu, Y., Blom, P.W.M., Damjanovic, D., Nielsen, M.M., de Leeuw, D.M.: The negative piezoelectric effect of the ferroelectric polymer poly(vinylidene fluoride). Nat. Mater. **15**, 78 (2015)

Chapter 4
Wearable Electronic Textiles from Nanostructured Piezoelectric Fibers

4.1 Introduction

PVDF base materials showed limited piezoelectricity and power output because their active material was based on ferroelectric polymers with high piezoelectric voltage constant but a low dielectric constant (ε ~ 30). One potential method to increase the dielectric constant of the PVDF polymers is to introduce high dielectric constant materials such as inorganic piezoelectric materials (i.e. barium titanate nanoparticles) into the PVDF matrix [1, 2]. However, inorganic piezoelectric materials, such as, lead zirconate titanate (PZT) [3], barium titanate (BT) [4], sodium niobate [5], lead magnesium niobate lead titanate [6] and zinc stannate (Zn_2SnO_4) [7] have high significant piezoelectric performance but their rigid nature limit their application in flexible self-powered devices [2]. Hybrid piezoelectric composites for energy harvesting based on barium titanate (BT) is attractive because of its high-dielectric-constant (ε ~ 2000), low cost, natural abundance, and environmental friendliness. However, overcoming the limited flexibility and durability of the BT still remains an unavoidable challenge to be addressed for optimization of its energy harvesting performance. Consequently, several research groups have introduced some structural strategies for piezoelectric polymer design to enhance the piezoelectric efficiency by incorporating inorganic piezoelectric nanostructures as an effective piezo electron pathway [8]. Using this strategy the voltage and current outputs of the P(VDF-TrFE) nanofibers could be enhanced up to 200% by adding BT nanoparticles into the polymer matrix [9]. It was also reported that flexible piezoelectric energy generators based on PVDF-HFP/BT composite film exhibited high electrical output up to ~75 V and ~15 μA [8]. Furthermore, a novel strategy was developed to improve the interface effect of PVDF/BT nanocomposites to enhance energy density from 6.5 to 9.01 J cm^{-3} [10] Nevertheless, such polymer based energy harvesters still has limitation of complex fabrication methods and low flexibility which could be problems to expand their applications in wearable electronics. To solve these problems, we explore a novel approach to develop high performance textiles-based energy harvesting devices as next generation wearable energy generators and sensors. Here, different variation of

F. Mokhtari, *Self-Powered Smart Fabrics for Wearable Technologies*, Springer Theses,
https://doi.org/10.1007/978-3-031-06481-4_4

textile designs from melt-spinning, knitting, weaving and braiding with the capability of low-cost, high production speed and high performance are developed. The high-performance PVDF piezoelectric nanocomposites fibers with and without BT nanoparticles were produced through a melt-spinning process. The effect of BT nanoparticles additions on the piezoelectric properties of hybrid PVDF/BT fibers with different ratio of the BT nanoparticles was investigated by fabricated these fibers into triaxial braided energy harversters, as previously reported [11].

The optimised PVDF/BT nanocomposite fibers were then used to create wearable woven and knitted energy harversters and strain sensors. The wearable energy harvesting textile fabrics were made from developed piezoelectric fibers (Fig. 4.1). A seamless feed-in process for integrating conductive fibers as electrodes with piezoelectric fibers was established to allow continuous fabrication of the energy harvesting textiles. As-fabricated knitted, triaxial braided and woven energy harvesting devices provide better mechanical properties (i.e. durability, flexibility and comfort) and piezoelectric performance in compare with PVDF fibers. The developed textile energy harvesting devices are durable, light and flexible and it is expected to be practical for wearable devices with high performance in near future.

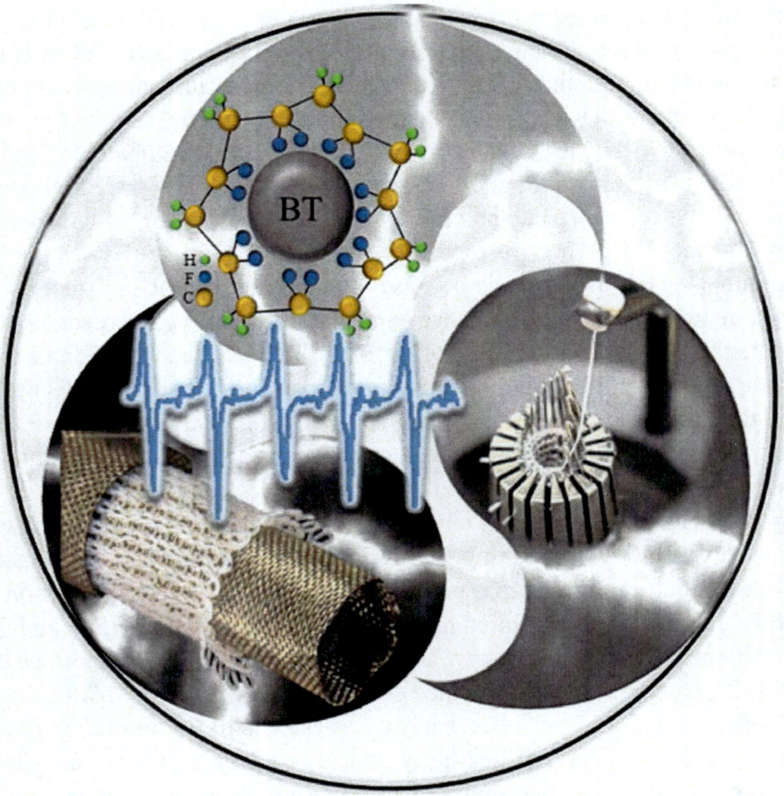

Fig. 4.1 Schematic illustration of the fabrication of the wearable knitted energy generator from nanostructured hybrid PVDF/BT fibers

4.2 Experimental

4.2.1 Fabrication of Wearable Energy Harvester

4.2.1.1 Knitted Wearable Energy Harvester

Knitted wearable sensors and energy generator based on as-prepared nanostructured PVDF and hybrid PVDF/BT fibers have been developed using circular knitting machine as explained in Sect. 2.2.9. As can be seen from Fig. 4.2a, as-prepared melt-spun piezoelectric fibers were suitable for direct use in a knitting machine and could sustain the applied mechanical stress and strain during the knitting process without breakage.

To assemble the knitted wearable energy harvester device (Fig. 4.2b, c), the commercially available woven conductive fabric as inner and outer electrodes with thickness of 80 μm was embedded inside and outside of the knitted structure (Fig. 4.2d). The use of the circular knitting would be able to provide protection against short circuiting by surrounding the inner electrode by the knitted structure. In addition, electrode bands could be fabricated into the knitted structure as needed and for example a wearable knitted device with two knitted electrodes into the structure has been developed.

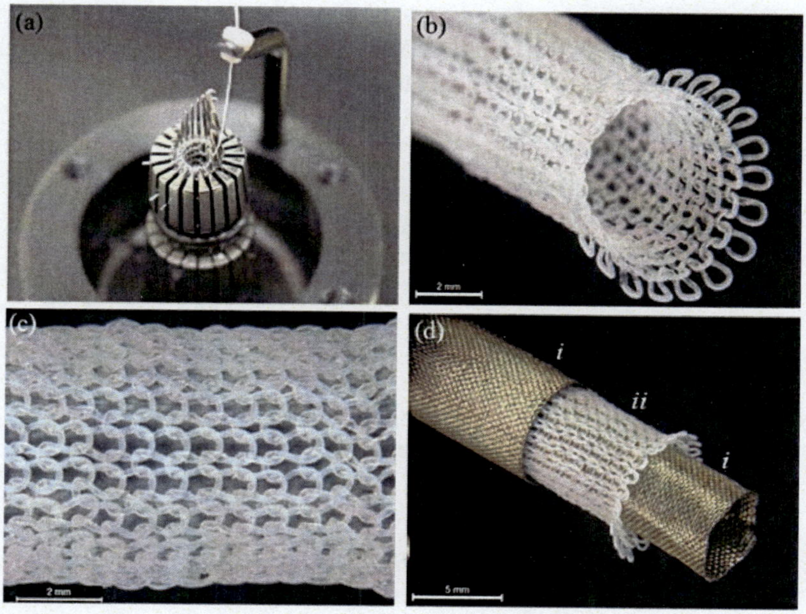

Fig. 4.2 **a** The process for producing a knitted structure by circular knitting machine, **b** circular knitted structure of PVDF/BT fiber, **c** magnified image of loops formation in knitted structure, **d** optical photograph of the circular knitted piezo generator (i) conductive woven fabric as electrodes and (ii) knitted PVDF/BT fiber as the middle layer

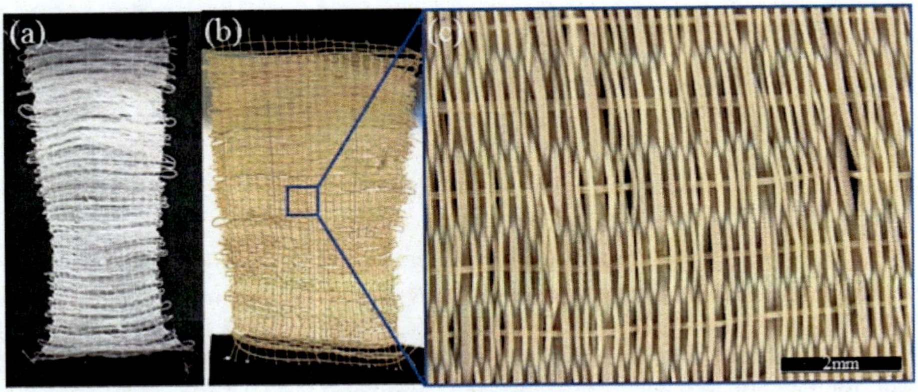

Fig. 4.3 Woven structure of: **a** pure PVDF fiber, **b** PVDF/BT fiber and **c** magnified image of PVDF/BT woven structure

4.2.1.2 Woven Wearable Energy Harvester

Wearable energy generator and sensors based on as-prepared PVDF and PVDF/BT nanocomposites fibers were developed through weaving process. The preparation was based on plain weave structure which is each weft yarn passes above and below (riser and sinker) the warp yarns repetitively so forming a simple cross pattern (Fig. 4.3). In the plain weave the short length of yarn intertwined between warp and weft yarn which produces a fabric with high density and consequently prevent short circuits between two electrodes. Moreover, the plain weave structure has a homogeneous surface to provide moderately constant electrical properties [12]. The thickness of the developed PVDF and PVDF/BT woven structure was 260 μm. The linear density of as-prepared PVDF and PVDF/BT woven structure were 0.057 and 0.029 g cm^{-2} respectively.

To assemble the woven wearable energy harvester device, the commercially available woven conductive fabric as electrodes with thickness of 80 μm was attached to the top and bottom of as-prepared fabric using a sewing machine (Fig. 4.4).

4.2.2 Morphology and Characterization of PVDF and PVDF/BT Nanocomposite

Scanning Electron Microscopy (SEM) was applied to reveal the morphological variations of PVDF/BT films and fibers prepared with different ratios of BT to PVDF. As can be seen from Fig. 4.5, an agglomerated structure occurred in the PVDF/BT nanocomposite film that were cast from solution and when the amount of BT nanoparticles was more than 10 wt% in the polymer matrix.

Both neat PVDF and hybrid PVDF/BT were prepared in a two-stage process involving film casting followed by melt spinning. Films were cast containing 0,

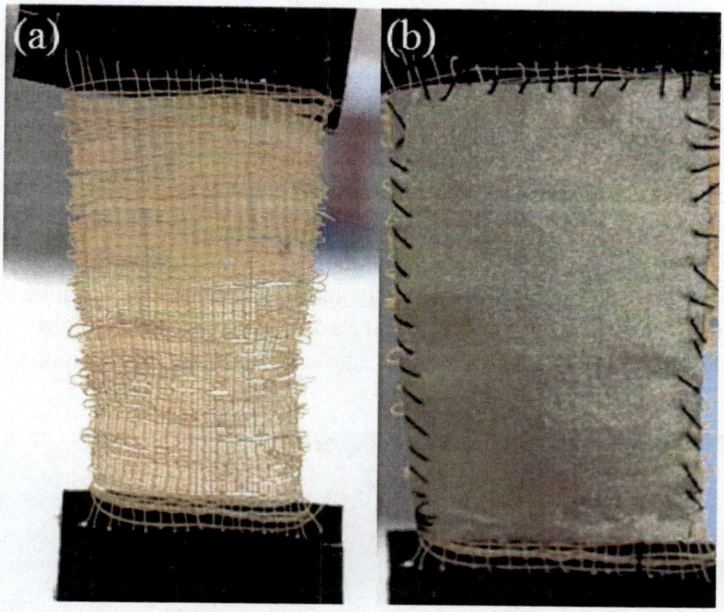

Fig. 4.4 **a** Plain woven PVDF/BT structure, **b** final structure for woven PVDF/BT filaments with two electrodes on top and bottom

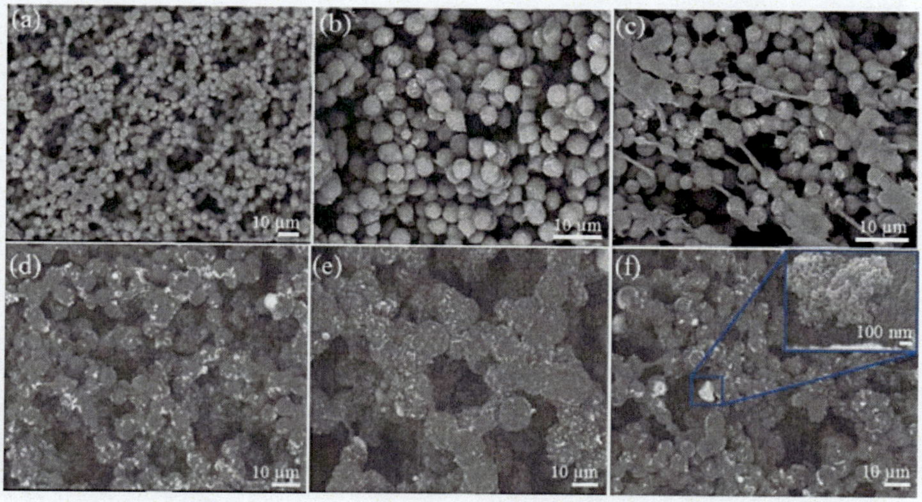

Fig. 4.5 SEM image of as-prepared cast films: **a** neat PVDF, **b** PVDF/BT (5 wt%), **c** PVDF/BT (10 wt%), **d** PVDF/BT (15 wt%), **e** PVDF/BT (20 wt%), **f** PVDF/BT (25 wt%)

5, 10 and 20 wt% of the BT nanoparticles. Due to agglomeration for PVDF/BT composites of 25 wt%, this composite did not study anymore. The as-prepared ground PVDF/BT nanocomposite film was fed into an extruder to aid homogenisation by shear forces and flow pressure. It was found that preparing nanocomposites with more than 10 wt% of BT was difficult due to the limited mobility of the polymer chains even at the molten state and this method was not able to disperse the aggregated BT particles within the polymer properly [13]. SEM micrographs of the as-spun PVDF and PVDF/BT nanocomposite fibers are shown in Fig. 4.6. As can be seen from the surface morphology and cross-section (Fig. 4.6a–d), both melt-spun PVDF and hybrid PVDF/BT fibers are very smooth and without any observable porosity or voids. SEM images of the cross-section of PVDF/BT nanocomposite fiber shows a very dense structure with the uniform distribution of nanoparticles throughout the fiber (Fig. 4.6b, c). The BT nanoparticles in the polymer matrix can be seen clearly in the fiber cross-section at the higher resolution (Fig. 4.6e). Figure 4.6f shows nanostructured BT nanoparticles. As is shown from elemental mapping analysis in

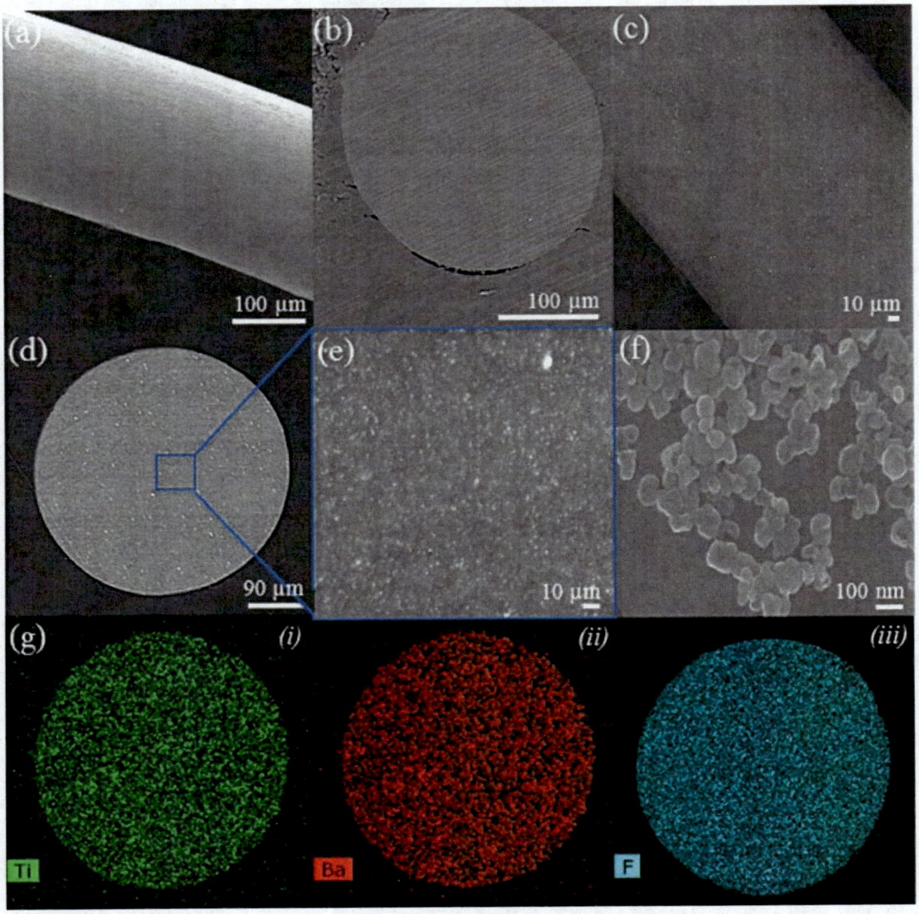

Fig. 4.6 SEM images of the as-spun PVDF and PVDF/BT fibers **a** surface of PVDF fiber, **b** cross section of PVDF fiber, **c** surface of PVDF/BT (10 wt%), **d** cross-section of PVDF/BT at low and **e** higher magnification. **f** Nanostructured BT nanoparticles. **g** Elemental mapping analysis performed on the hybrid PVDF/BT (10 wt%) fiber cross-section area (i) the location of titanate (Ti green dots), (ii) barium (Ba red dots) and (iii) Fluoride (F blue dots)

Fig. 4.6g(i, ii, iii), the hybrid nanoscopic structure of the melt-spun PVDF/BT was confirmed through the overlapped Ti, Ba, F mapping images (Fig. 4.6g). The phase transformations of the as-prepared PVDF and hybrid PVDF/BT fibers following different amounts of additive (BT) was scrutinised using FTIR, DSC, TGA and X-ray diffraction. Piezoelectric properties of PVDF fibers can be enhanced with a higher fraction of β phase. The FTIR results (Fig. 4.7a) indicated that the ratio of the BT nanoparticles in the polymer matrix could affect β phase formation in the PVDF/BT nanocomposite fiber (Fig. 4.7a). The vibrational bands at 764, 976 cm^{-1} are attributed to the non-polar α phase, whereas the characteristic peaks at 841 and 1276 cm^{-1} correspond to the electroactive β phase [14]. The relative fraction of the β-phase in a sample containing just α and β-PVDF was calculated from Eq. (3.1). The characteristic peaks of the α phase severely decrease in the PVDF/BT fiber in compared with the pure PVDF which signify that adding BT nanoparticles is a highly efficient method of inducing fraction of the polar β phase.

The variation of β phase fractions for as-prepared PVDF and PVDF/BT nanocomposite fibers is shown in Figs. 4.7b and 4.8 for nanocomposite films.

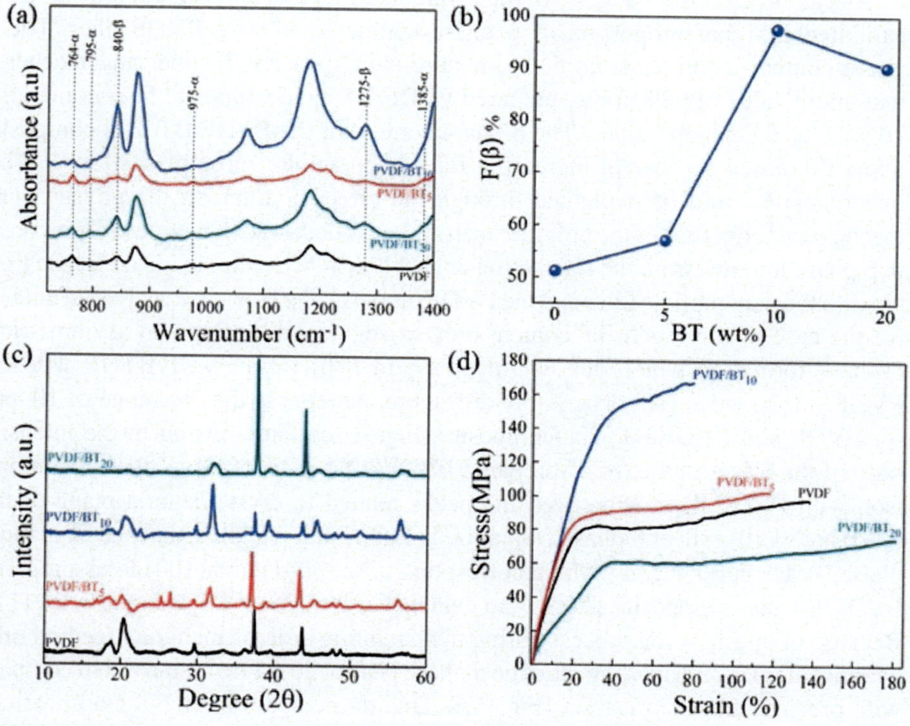

Fig. 4.7 Characterisation of the melt-spun PVDF and PVDF/BT nanocomposite fibers with different percentage of BT nanoparticles (5, 10, 20 wt%): **a** FTIR spectra of PVDF and hybrid PVDF/BT fibers. **b** The variation of the calculated β phase contents of the PVDF fiber as a function of the percentage of BT nanoparticles (wt%). **c** X-ray diffraction patterns for PVDF and hybrid PVDF/BT fibers, and **d** Stress–strain curves obtained from tensile tests of PVDF and PVDF/BT nanocomposite fibers

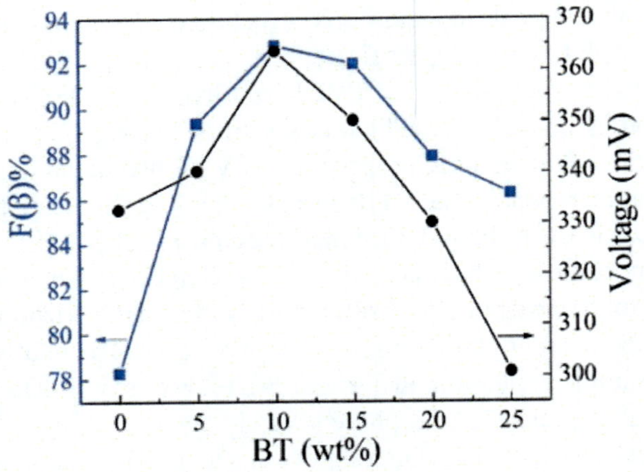

Fig. 4.8 Variation of β phase contents and output voltage of the as-prepared PVDF/BT naoncomposites films as function of percentage of the BT nanoparticles

As can be seen from Fig. 4.7b, the as-prepared PVDF/BT nanocomposite fibers exhibited a higher proportion of β phase compared to pure PVDF fiber. The β phase content increases from 51% for pure melt-spun PVDF fiber and reaches a maximum value of 98% for as-prepared PVDF/BT nanocomposite fiber containing 10 wt% of BT nanoparticles. The β phase content of the PVDF/BT nanocomposite fibers decreased because of increasing BT percentage for more than 10 wt%. This phenomenon could be explained through the effect of filler on the physical and mechanical properties of the polymer matrix. It was observed that noticeable increase in β phase initiates from the interaction enhancement between the local electric field close to the nanoparticle filler and the PVDF dipoles [15]. However, increased defects for the case of too much BT content prevent segmental motion and asymmetrical β-phase formation which due to a decrease in $F(\beta)$ from PVDF/BT(10 wt%) to PVDF/BT(20 wt%) [16]. The X-ray diffraction patterns in the 2θ range of 10–60° for PVDF and PVDF/BT nanocomposite (Fig. 4.7c) show raising in the intensity ratio of the β to α phase for as-prepared PVDF/BT nanocomposite fiber in comparison with PVDF fiber. Moreover, the peaks related to crystallization plane of the (020) and (110) reflect β phase formation at 20.6°, indicate the existence of α and β phase. Peaks at 2θ ~ 22.2° which correspond to the (001) and (100) in as-prepared PVDF/BT nanocomposite fibers is an evident for tetragonal BT nanoparticles [17]. Because of optimisation process during meltspinning process including feeding rate, heating, die size and drawing ratio the α phase peak at 2θ ~ 17.6° removed in compare with previous fiber fabrication (Fig. 3.6). Due to the addition of BT modifications in the crystallinity of PVDF have been investigated on the basis of the obtained X-ray pattern. As the concentration of BT increases, the intensity of standard peaks in the X-ray pattern of PVDF decreases and slightly displaces towards shorter angle. These peak shifts indicates the existence of specific interaction between the different phases of PVDF and BT [18]. In addition, thermal analysis of the PVDF and hybrid

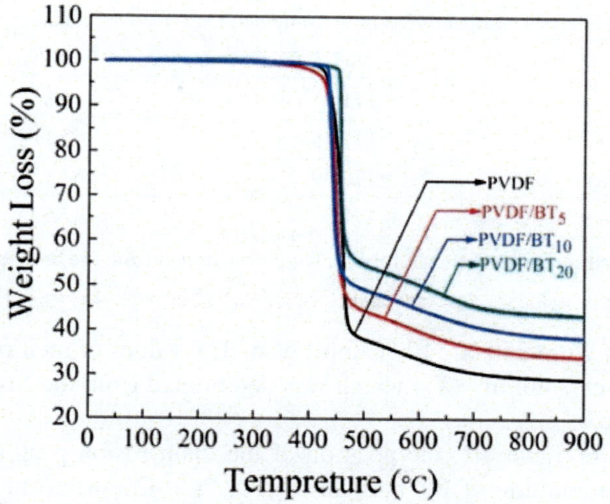

Fig. 4.9 TGA curves of PVDF and PVDF/BT nanocomposite fibers with different amount of the BT nanoparticles

PVDF/BT fibers specified that the crystalline structure formation in the as-prepared piezoelectric fibers are 41, 62, 65 and 48% for the as-prepared PVDF, PVDF/BT_5, PVDF/BT_{10} and PVDF/BT_{20} fibers, respectively (Fig. 4.9).

The thermal behaviours of the PVDF/BT nanocomposites were investigated using a differential scanning calorimeter (DSC) (Fig. 4.10). Four samples of each nanocomposite material, about 5 mg, were directly cut from filaments with different concentration of BT. Melting temperature and crystallinity were determined using the first

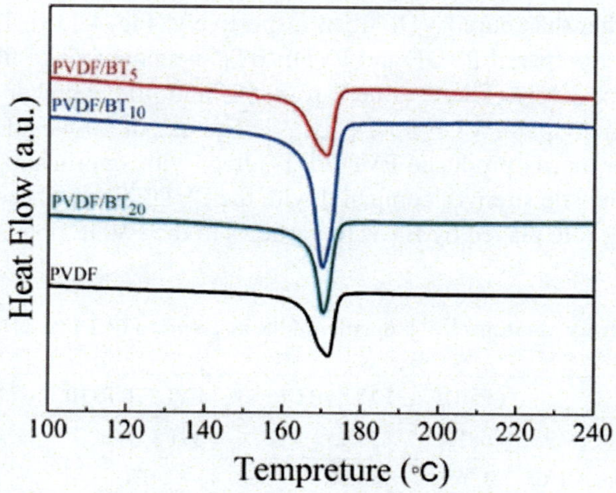

Fig. 4.10 DSC patterns for PVDF and PVDF nanocomposite fibers with different amount of BT nanoparticles

Table 4.1 Characteristic temperatures values obtained from the DSC data

Samples	Tm (°C)	Xc (%)
PVDF	171.7	40.83
PVDF/BT (5%)	171.35	62.11
PVDF/BT (10%)	170.39	64.85
PVDF/BT (20%)	170.5	47.5

Melting temperature Tm and crystallinity Xc were determined from the first heating DSC curves

heating (from 25 to 250 °C, 10 °C/min) data. The values of melting (T_m) temperatures and the crystallinity (X_c) which was determined from the fusion enthalpy are listed in Table 4.1.

As shown in Table 4.1, the position of the main fusion peak (T_m) was hardly altered by the addition of BT particles (~171 °C). Compared to the pure PVDF, the nanocomposite samples exhibited higher crystallinity regardless of BT content. The crystallinity firstly increased with increasing BT content up to 10 wt%, which is due to the nucleating effect of BT particles, then decreased at higher BT content (20 wt%), resulting from the reduced chain mobility with a high filler content. The as-prepared PVDF and PVDF/BT nanocomposite fibers have also been investigated to evaluate the effect of BT nanoparticles on the mechanical properties. During melt spinning and cold drawing process fiber are continually under tension which brings tenacity and elasticity for the final fibres [19].

A comparison of the stress-strain curves for the various prepared fibers is given in Fig. 4.8d. As can been seen from Fig. 4.11 and Table 4.2, the ultimate tensile strength and elastic modulus of the prepared fibers have significantly increased for PVDF/BT_5 and PVDF/BT_{10} compared to PVDF fiber. Young's modulus and tensile strength of the as-prepared PVDF/BT nanocomposite fiber with 10 wt% BT is 130% and 170% higher than pure PVDF fiber, respectively (Fig. 4.11a). The elongation at break of the as-prepared PVDF and PVDF/BT_{10} nanocomposite fibers were 137% and 80%, respectively. These values are 685% and 400% higher than previously reported for melt-spun PVDF fiber [11, 19]. The use of 20 wt% BT nanoparticles in the polymer matrix produced PVDF/BT_{20} fiber with significantly lower Young's modulus and tensile strength compared with the PVDF fiber. These results confirm the reinforcing role played by BT nanoparticles in the PVDF fibers.

Table 4.2 Summery of mechanical properties of the as-prepared PVDF and PVDF/BT nanocomposite fibers

Sample	PVDF	PVDF/BT (5%)	PVDF/BT (10%)	PVDF/BT (20%)
Young's modulus (MPa)	681.4	839.1	891.1	411.8
Elongation at break (%)	136.88	117.88	79.338	180.36
Tensile strength (MPa)	97.93	102.31	167.461	74.9

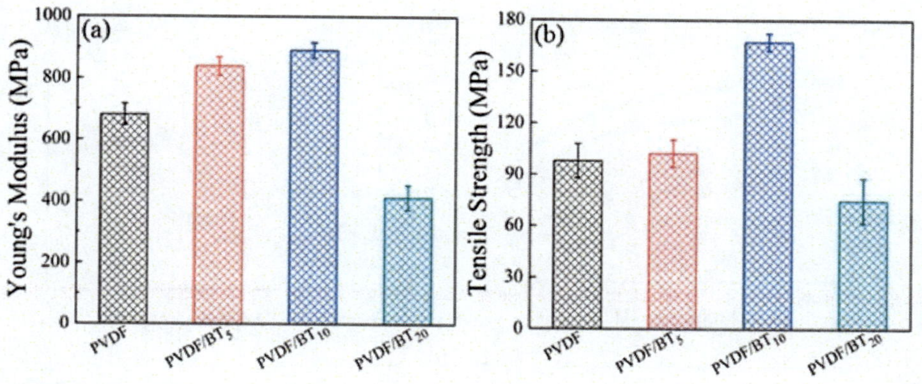

Fig. 4.11 Young's modulus and tensile strength of PVDF and PVDF/BT nanocomposite fibers at different BT concentration

However, the decrease in mechanical properties of the PVDF/BT_{20} fiber with addition of more than 10 wt% BT nanoparticles may be due to aggregation of nanoparticles and/or phase separation of the polymer (Fig. 4.5). This phenomena can be due to the stress-concentration or low adhesion at the phase interface that would lead to the lower tensile strength and modulus [20–22].

4.3 Wearable Energy Generator and Sensor Performance

Initial evaluation of piezoelectric energy harvesting performance of the hybrid fibers used the triaxial braided textile structure described in our previously reported results [11]. The dielectric constant (ε_r) and dielectric loss (tan δ) of the as-prepared braided wearable energy generators were measured at room temperature in a frequency range up to 10^6 Hz as shown in Fig. 4.12a, b. It was reported [16, 23, 24] that piezoelectric properties of PVDF nanocomposite polymer could be significantly improved for the piezoelectric polymer with high dielectric constant and low dielectric loss due to enhancing its electroactive β phase. The dielectric constant influences the performance of a piezoelectric power generator. Higher dielectric constants in harvester systems lead to larger power output [24]. It should be also noted that the piezoelectric coefficient is linearly proportional to the dielectric constant (ε) of the piezoelectric materials, i.e., $d_{33} \sim \varepsilon\ P_r$, where P_r is the remnant polarization [25]. The dielectric constant of the braids prepared with the hybrid fibers was decreased with increasing frequency for all samples. However, the dielectric constant values were higher in braids made with higher BT content fibers due to the large dielectric constant of BT (Fig. 4.12a). The gradual decrease in dielectric constant when measured at higher frequencies can be attributed to the reduction in dipole mobility where the dipoles are not sufficiently mobile to displace to the same extent when the frequency of the applied electric field exceeds the relaxation frequency [26]. On the other hand, the

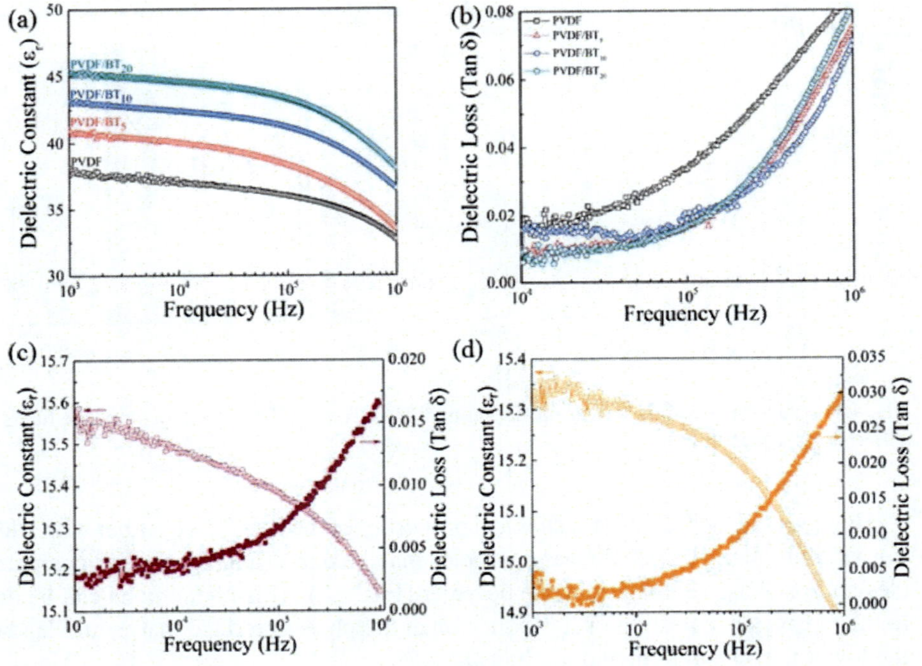

Fig. 4.12 The dielectric properties of the as-prepared textile energy generator as a function of frequency: **a** dielectric constant of the braided energy generator; **b** loss tangent of the braided energy generator; **c**, **d** dielectric constant and dielectric loss as a function of frequency for woven and knitted wearable energy generators based on PVDF/BT_{10} nanocomposite fiber, respectively

dielectric loss increased only slightly with increasing BT contents. The dispersion of the BT nanoparticles into the polymer matrix increases the filler/polymer interfacial area and formation of the β phase fraction and consequently increase dielectric properties and decrease loss tangent (Fig. 4.12c, d). Electroactive β phase formation can be explained by the surface charge/dipole interaction occurring between the BT nanoparticles and PVDF chains [27]. The BT nanoparticles surface charge contribute significantly in the electroactive β phase nucleation procedure [28]. This step-like decrease in dielectric constant with frequency may be explained by the Maxwell–Wagner–Sillars interfacial polarization mechanism [29]. The voltage output of the braided PVDF and PVDF/BT nanocomposite fibers occurring as a result of mechanical deformation caused by repeated impact is shown in Figs. 4.8 and 4.13. Mechanical stimulation of the braid occurred by compressing the braid between a solid flat surface and platen that was driven vertically by an oscillating cam attached to a reciprocating motor. The amplitude and frequency of the cyclic compression were initially kept constant to investigate the effect of fiber composition on the piezoelectric output of the braided energy generator. An example of the developed braided energy generator device is shown in Fig. 4.13a. As can be seen from Fig. 4.13c, the voltage output of the triaxial braided energy generator was significantly improved due to synergistic effects of piezoelectric BT nanoparticles incorporated into the PVDF piezoelectric polymer. As-spun PVDF fiber generated a maximum voltage output of 480 mV, while the voltage output of the braided PVDF/BT_{10} fiber was found to be 1100 mV. This value is 229% and 250% higher, respectively, than PVDF fiber or previously reported for melt-spun hybrid PVDF/BT fiber with the same 10 wt% content of BT nanoparticles [11, 30]. The voltage output of the braided PVDF/BT_{20} with more than 10 wt% BT nanoparticles generated a maximum voltage output of 895 mV which is much lower compared to the braided PVDF/BT_{10} fiber. The voltage output can be estimated using the following expression:

$$V = g_{33}\varepsilon YD \tag{4.1}$$

which voltage output (V), strain (ε), Young's modulus (Y), piezoelectric voltage constant (g_{33}) and the fiber diameter (D) have the above relation. The D and ε are similar in all the as-prepared hybrid PVDF/BT fibers, and the Y is higher in PVDF/BT_{10} than that in PVDF/BT_{20} nanocomposite fibers (~891 and 412 MPa, respectively).

The measured V is higher in the as-prepared PVDF/BT_{10} fiber than that in PVDF/BT_{20} fiber, but the combined analysis indicates the g_{33} is much larger in PVDF/BT_{20} than in PVDF/BT_{10} meaning the piezoresponse in PVDF/BT_{20} fibers is more sensitive to external stress [16]. Adding piezoelectric nanoparticles to the polymer matrix can negatively affect the piezoelectric performance of the composite because of the positive and negative piezoelectric co-efficient of BT nanoparticle and PVDF polymer and cancelling their effect [23, 24, 31].

The voltage output of the braided PVDF/BT_{10} energy generator device under cyclic impact was further investigated. The generated positive and negative pulse signals corresponding to the pressing and releasing process during a cyclic impact is shown in Fig. 4.13b which was selected and magnified from Fig. 4.13c in the coloured background region. A graph of instantaneous power at different load resistances is shown in Fig. 4.13d. The power generated by the as-prepared PVDF and PVDF/BT nanocomposites fibers was calculated through Eq. (4.2):

$$P = \frac{1}{T}\int \frac{U^2(t)}{R}dt \tag{4.2}$$

where U(t) is the real-time voltage integrated over time t, R is the external load resistance, and T is the period of load application [32]. With further increasing load resistance, the voltage starts to increase up to 3 V at the load resistance of 1000 kΩ. As a power source, the developed triaxial braided PVDF/BT_{10} piezo fibers reaches the maximum power output of ~0.21 μW when the load resistance equals to 400 kΩ. While the maximum power output and the load resistance for Pure PVDF fibers are ~0.16 μW and 900 kΩ respectively. Although fibers output impedance based on the measured voltage and current output is ~1 MΩ but it is covered with internal resistance of the Keithley which was used for measuring these data.

The amount for the load resistance is in the range according to the latest researches [33–35] which depends on the filler type and sample thickness. To demonstrate the

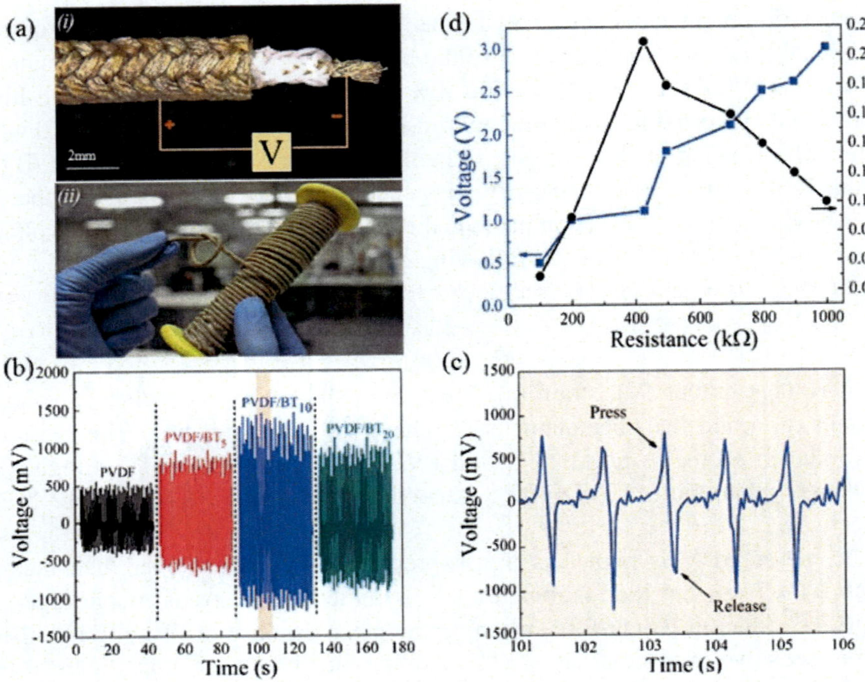

Fig. 4.13 **a** Digital photograph of triaxial braided piezo fibers: (i) illustration of energy harvesting mechanism, (ii) a bobbin of flexible triaxial braided energy harvester, **b** effect of BT nanoparticle content (wt%) on open circuit voltage of the as-prepared triaxial braided piezo fiber energy generators, **c** the magnified image of the open circuit voltage in the region of 101–106 s for the braided PVDF/BT_{10} nanocomposite fiber, **d** the voltage output and power dependence on the load resistance of the braided PVDF/BT_{10} which was obtained at an impact pressure of 0.0031 MPa

feasibility of harvesting energy using the flexible triaxial braided piezo nanocomposite fibers a bridge rectifier (4 diodes of 1N5817) was placed in the circuit to feed a different capacitor under mechanical pressures (Fig. 4.14a). While this was the more conventional approach adopted for this development stage, one can find a higher efficient circuit by Mirvakili et al. [36] Harvesting energy has a close relationship with the diode characteristics, and a diode with low reverse leakage current is favourable so that the Schottky diode 1N5817 is one the best options at this stage of development. For future, and more extreme target conditions, other more suitable options for higher frequency rates and lower drop voltages are recommended, e.g. Avago HSMS-285 C or SDM03U40.

Charging performance measurements were carried out upon mechanical force on the triaxial braided piezo nanocomposite fibers (PVDF/BT_{10}) in a RC circuit (Fig. 4.14b) using different capacitors (0.68, 2.2, 10 and 22 μF). It was observed that the build-up voltage increases exponentially and reaches a steady state as shown in Fig. 4.14c. In order to calculate the voltage across the capacitor (as shown in Fig. 4.14d) during charging capacitor, Eq. (4.3) was used which related to RC circuit charging:

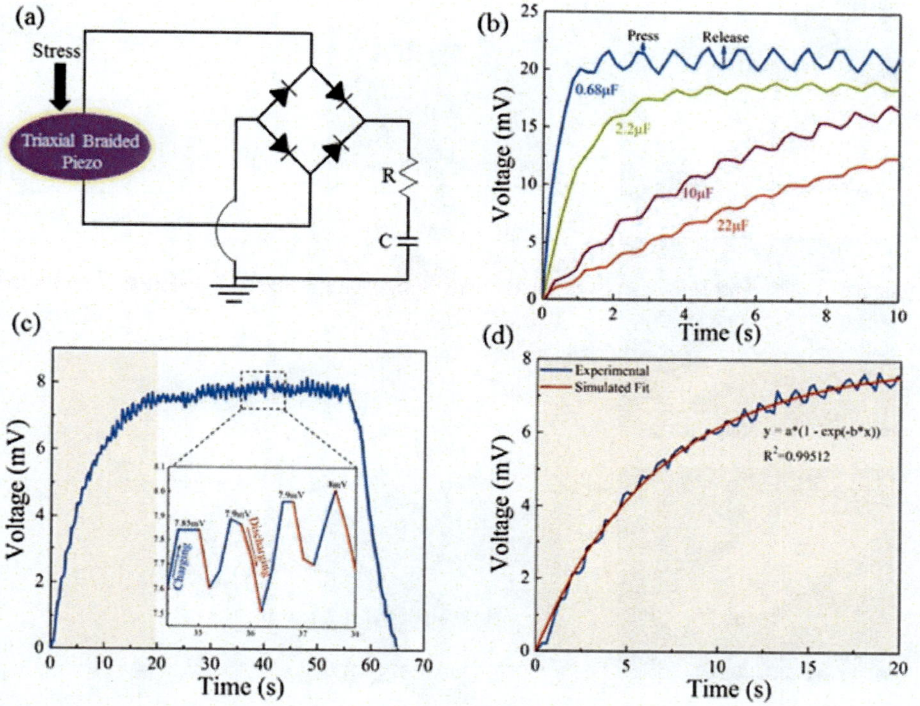

Fig. 4.14 **a** Equivalent of the full wave rectifier circuit for charging a capacitor with resistance 430 kΩ. **b** Charging of the different capacitors by the power generated from the four parallel connections of the braided PVDF/BT_{10} under periodic impact pressure. **c** Charging-discharging (V − t) graph of one triaxial braided piezo PVDF/BT_{10} structure across capacitor 10 μF and **d** Charging capacitor in experimental and simulated fit for magnified part of charging graph C

$$V_c(t) = V\left(1 - e^{\frac{-t}{RC}}\right) \tag{4.3}$$

where C is the capacitance of the capacitor, R is the resistance in RC circuit. The time constant ($\tau = RC$) was obtained from an exponential fit to the experimental data as shown in Fig. 4.14d. A slight difference between the experimental data and the fitted data may be due to the power consumption by the measuring unit present in the device during the measurements.

The comparison of braided energy generators prepared with different hybrid fibers indicated that the PVDF/BT_{10} composition provided optimal performance and these fibers were then used to prepare knitted and woven textiles, as shown in Fig. 4.15. The storage energy calculated for different as-developed energy harvesters in the capacitor was determined using capacitor potential energy formula:

$$E = \frac{1}{2}CV^2 \tag{4.4}$$

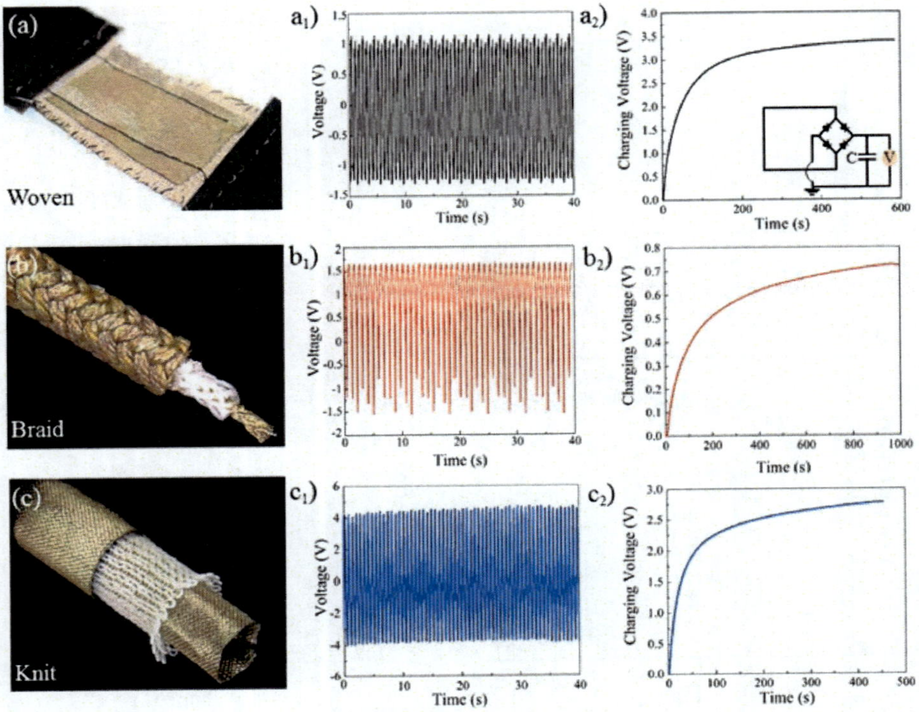

Fig. 4.15 The voltage output performance and the voltage profile of the charged capacitor (10 μF) using as-prepared wearable energy harvester based on PVDF/BT10 fibers: **a** woven energy generator, (a_1) voltage output versus time for woven generator (a_2) charging voltage versus time for woven generator, **b** triaxial braided energy generator (b_1) voltage output versus time for braided generator, (b_2) charging voltage versus time for braided generator **c** circular knitted energy generator, (c_1) voltage output versus time for knitted generator, (c_2) charging voltage versus time for knitted generator

where C is the capacitor capacitance, V is the charging voltage across the capacitor under steady state condition at a definite time (t) [37]. The energy storage based on the as-fabricated textiles energy harvesters (woven, triaxial braided and circular knitted structures in Fig. 4.15) in 10 μF capacitor were found to be 61.25, 2.81 and 36.45 μJ, respectively. The energy conversion efficiency of the as-developed triaxial braided PVDF/BT_{10} energy generator can be estimated as:

$$\eta = \frac{E_e}{E_s} \times 100 \tag{4.5}$$

the electric energy generated E_e from the device can be obtained by $E_e = \int VI\,dt$, where V and I are the measured output voltage and current density of one cycle time t. The mechanical strain energy E_s stored in the device is expressed as $E_s = 1/2YAL\varepsilon^2$, where A is the active area, L is the original thickness, ε is the strain, and Y is the Young's modulus of the material [38, 39]. The energy conversion efficiency of the as-developed triaxial braided, circular knitted and woven energy generators are calculated as 27%, 29% and 40%, respectively. The obtained results suggested that the developed textiles energy harvester would establish the viability of such wearable piezoelectric energy generators in real life applications. The knitted PVDF/BT_{10} energy generator was able to charge a 10 μF capacitor in just 400 s under periodic impact and relaxation (Fig. 4.15c_2). The obtained results are very promising compared to previous reported systems (Table 4.3). The circular knitted structure enables the ready integration of electrodes into the triaxial structure (inner and outer electrodes) which could enhance the collection of charge and energy conversion. The force sensitivity and power output of the wearable energy harvesters based on PVDF/BT_{10} textile structure were compared, as shown in Fig. 4.16a. The sensitivity of the wearable energy generators were assessed by a ratio of voltage output to the applied force when the textiles were subjected to compress using repeated impact as explained in Sect. 2.2.10.

The voltage output of the piezoelectric textiles were proportional to the applied force. It was found that the force sensitivity of the textile energy harvesters were 3, 4, and 10 V/N for woven, braided and knitted energy generators, respectively. The results revealed that the knitted energy generators exhibited an almost 6.3-fold increase in the value (10 V/N) compared to a recent report [11]. The power output of the wearable energy harvesters were found to be 36.2, 38.8 and 87 μW cm^{-3} under a periodic compression for woven, braided and knitted energy generators, respectively. The power output density of the woven wearable energy generator was significantly enhanced and it was 294 and 4578% higher than previous report for braided and woven piezoelectric energy generators, respectively [11, 19]. (Table 4.3).

The novelty of circular knitting and braiding techniques for piezoelectric fibers is their packaging structures which interwound fibers and due to improved durability and flexibility. The experimental results in our previous work [11] confirmed the stability performance of the triaxial braided piezoelectric fibers in the bending test during 1000 cycles to a maximum compression strain of 50% at 0.6 Hz with no change in its performance. The obtained results suggested that developed wearable energy generators would be able to charge the capacitors to a certain voltage under a specific time for the self-powered electronic devices. Consequently, the integration of the

Table 4.3 Comparison of power density of our flexible textile with other reported generators earlier

Generator	Output voltage (V)	Power density (μW cm^{-3})	References
PVDF/AlO-rGO	36 V	27.97	[37]
BaTiO3 nanotubes	5.5 V	64	[40]
PVDF/DNA	20 V	11	[41]
ZnO NWs/PVDF	0.2 V	2	[42]
$BaTi_2O_5$/PVDF	24 V	27.4	[43]
PVDF	380 mV	29.62	[11]
Knitted PVDF/BT_{10}	4 V	87 μWcm^{-3}	This work

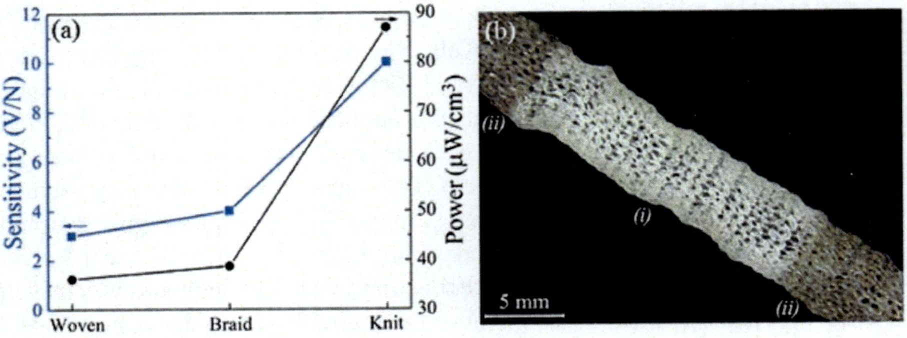

Fig. 4.16 **a** Comparing sensitivity and power output for as-prepared textiles energy generators, **b** wearable energy generator device based on integration of the seamless electrodes, (i) knitted PVDF/BT_{10} fiber (ii) knitted seamless integrating silver coated nylon yarn as electrodes

developed wearable energy generator with energy storage device (i.e. rechargeable battery or capacitors) can be of great potential for practical applications, including development of self-powered wireless sensors within garments and monitor the status of human health.

In addition, the developed wearable energy generator prototype device was integrated into a garment for charging a rechargeable battery or capacitor. Figure 4.17 shows the performance of the energy generator garment for biomechanical energy harvesting and storage during walking and/or running. The voltage output of the wearable energy generator could be tailored in the range of 300 (Walking) to 1000 mV (Running) as shown in Fig. 4.17b. As can be seen from Fig. 4.17c, the wearable energy generator could increase the voltage of the storage capacitor from 0 to 25 mV in 20 s so that the 10 μF capacitor was fully charged after approximately 25 steps at a running frequency of 1.2 Hz. This fast charging rate is 4 and 6 times faster compared to previously reported for the nanofiber piezoelectric PVDF/BT energy generator and hybridized energy conversion and storage based on PVDF film [9, 44].

The choice of 10 μF was sufficient to demonstrate full charging of a capacitor at this stage of development, with minimum input (25 steps), a higher storage (>500 μF) is recommended for applications outside the laboratory, e.g. sports training demonstrators. In such cases, it is also recommended to increase the voltage output of the system through a selection of approaches, e.g. by implementing voltage multiplication, parallel charge-serial discharge techniques. We have explored the potential applications of the developed wearable energy harvester as movement sensors.

As demonstration examples, both wearable and portable textile sensors have been developed as shown in Figs. 4.18 and 4.19. Wearable movement sensors are useful for real-time precise healthcare applications. Here a knee sleeve prototype device from a woven PVDF/BT_{10} nanocomposite fibers was established to support personal recovery after an injury (Fig. 4.18). A wireless data acquisition board could be assigned to the developed knee sleeve prototype device for data transmission of knee flexion resulting from tension and bending of the woven textile sensor. As can be

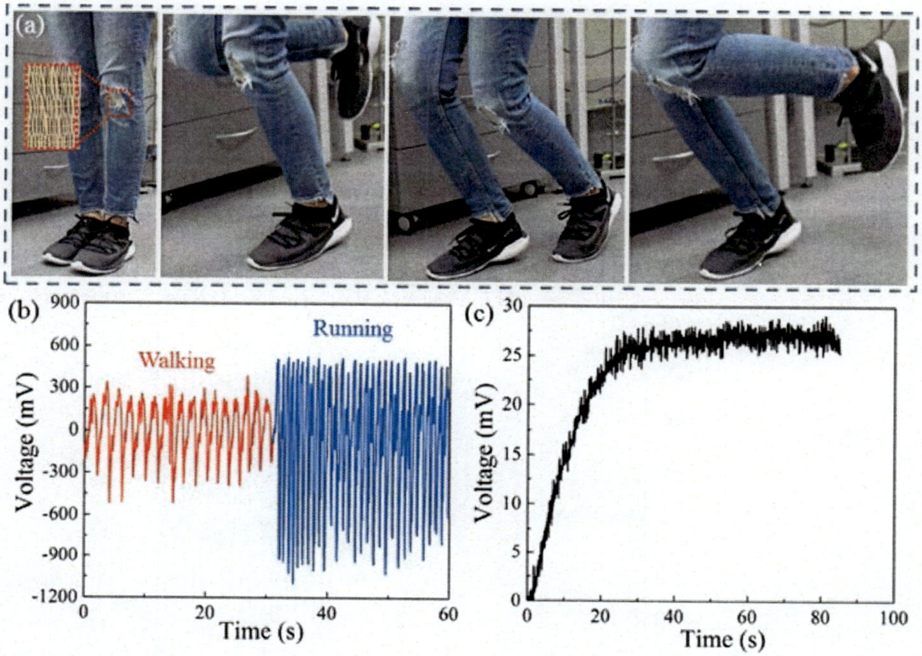

Fig. 4.17 **a** Photograph of the developed wearable and portable energy generator based on woven PVDF/BT10 piezoelectric fibers, **b** generated voltage from joint bending during walking and running and **c** charging 10 μF capacitor during 25 steps

seen from Fig. 4.18a, b, the developed knee sleeve prototype device generated a voltage output that varied with knee bending angle from the initial (zero bending Fig. 4.18a), bent (45° bending Fig. 4.18a_1) and bent (90° bending Fig. 4.18a_2). A voltage divider used to connect an expansion board (AnEx board) and a portable wireless device. The AnEx board can be attached to the knee sleeve and linked to it by an external connector. The collected data can be transferred via Bluetooth to a laptop for signal analysis. To demonstrate potential applications of the developed circular knitted piezoelectric strain sensor for the detection of human and/or industrial activities, knitted wearable piezoelectric fibers were assembled into two wearable sensors with the required structures to track these activities as shown in Fig. 4.19. The sensing performance of the developed knitted piezoelectric PVDF/BT_{10} fibers based on assembling core-sheath structure has been evaluated as shown in Fig. 4.19a. The performance of the potential real-life applications of the developed wearable sensor under biomechanical pressure of periodic finger pressure and relaxation was demonstrated in Fig. 4.19a_1–a_2. Furthermore, the versatility of circular knitted piezoelectric PVDF/BT10 sensors was demonstrated by knitting the textile with seamlessly integrated electrodes (Figs. 4.16b and 4.19b). Sequentially feeding silver coated nylon, followed by PVDF/BT_{10} fibers and then again silver coated nylon into the knitting machine produced the textile structure shown in Fig. 4.19b. This textile was coated with silicon resin to enhance its durability as well as creating new applications such as

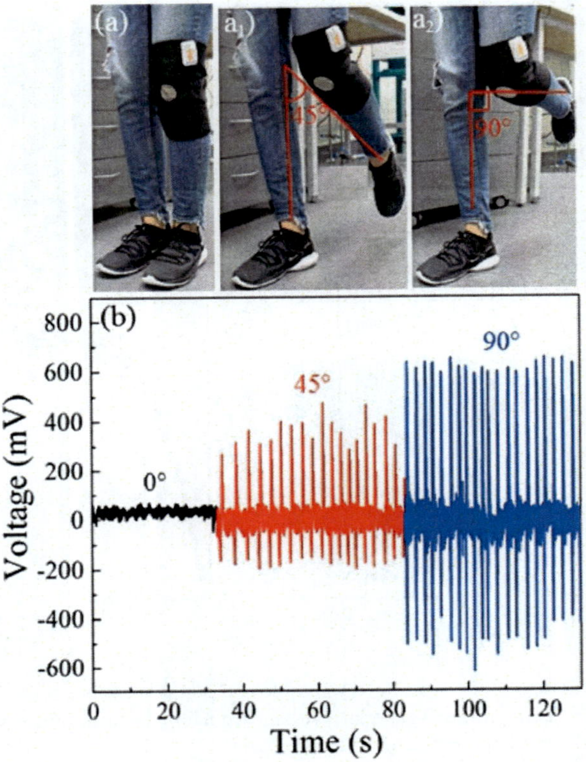

Fig. 4.18 Photograph of the developed portable knee sleeve based on wearable woven PVDF/BT10 piezoelectric sensor: **a** wearable strain sensor at initial zero bending, (a_1) bent at 45°, (a_2) bent at 90°. **b** The generated voltage associated with repeated knee bending and unbending to the maximum angle of 0°, 45° or 90°, as indicated

hydraulic and/or pneumatic pressure sensors. The performance of the silicon coated knitted piezoelectric sensor under biomechanical pressure of periodic finger pressure and relaxation was shown in Fig. 4.19b_1–b_2. The response time of the silicone coated knitted sensor (Fig. 4.19b_2) was significantly improved (2.5 s) as compared to core-sheath knitted structure (5 s) in Fig. 4.19a_2. In addition, the capability of the silicone coated knitted piezoelectric structure as hydraulic and/or pneumatic pressure sensors was demonstrated in Fig. 4.19c, d. As can be seen from Fig. 4.19c–c_1, the developed knitted sensor was subjected under hydraulic pressure where 2 ml water was pumped into the developed hollow structure (5 mm diameter and length of 20 mm) and it exhibited very fast response time (~2 s) with higher voltage output compared to other (Fig. 4.19a_2, b_2). Figure 4.19d–d_2 shows a knitted sensor under pneumatic pressure of 20 kPa. The development of silicon coated circular knitted piezoelectric structure capable of sensing pressure is of great importance for various application such as heart rate detection, pressure monitoring, strain gauges, robots, etc. due to its ability to response to bending, twisting, and compression motion. More importantly,

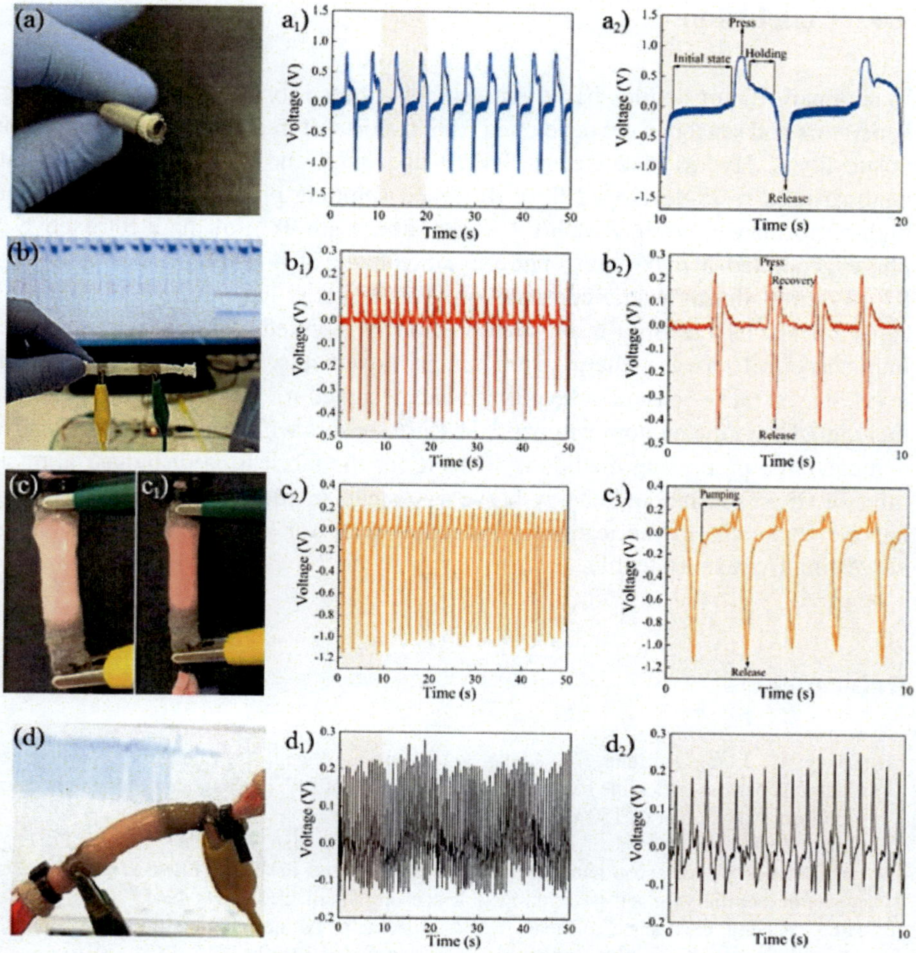

Fig. 4.19 **a** Photograph of the developed circular knitted piezo sensors: **a** assembled knitted piezo sensors with core-shell electrodes under hand pressure, (a_1) effect of pressure cycle on voltage output, (a_2) response time (voltage output vs. time) for magnified image (a_1), **b** developed silicon coated knitted piezo sensor with seamless integrating electrodes, (b_1) effect of pressure cycle on voltage output, (b_2) response time (voltage output vs. time) for magnified image b_1, **c** (c_1) the knitted piezo sensor before and after pumping water, respectively, (c_2) effect of pumping and release water (pressure cycle) on output voltage and (c_3) response time (voltage output vs. time) for magnified image c_2, **d** the knitted piezo sensor during pneumatic test under pressure of 20 kPa, (d_1) effect of pumping and release air (pressure cycle) on voltage output and (d_2) response time (voltage output vs. time) for magnified image d_1

developed processing method is scalable for the fabrication of industrial quantities of strain sensing and smart textiles.

4.4 Conclusion

In summary, smart textiles based on wearable knitted, braided and woven energy generators and sensors were developed from nanostructured piezoelectric nanocomposite fibers. Hybrid piezoelectric PVDF fiber with and without barium titanate nanoparticles were developed through a melt-spinning process. It was found that high-performance hybrid PVDF/BT_{10} piezofiber with 98% of the electroactive β-phase generated a maximum open circuit voltage of 4 V and a power density 87 μW cm^{-3} during cyclic compression. These energy generators could charge a capacitor (10 μF) 6 times faster than previously reported and could be integrated into garments to harvest energy from human movements. In addition, the wearable textile-based piezo sensors were utilized and designed for various sensing response including hydraulic and/or pneumatic pressure sensors with tunable sensitivity. The demonstrated processing method is scalable for the fabrication of industrial quantities of strain sensing and energy harvesting smart textiles. The presented method has the advantage of fabrication in industrial scale for strain sensing and energy harvesting from smart textile.

References

1. Zhang, G., Liao, Q., Zhang, Z., Liang, Q., Zhao, Y., Zheng, X., Zhang, Y.: Novel piezoelectric paper-based flexible nanogenerators composed of $BATIO_3$ nanoparticles and bacterial cellulose. Adv. Sci. **3**(2), 1500257 (2016)
2. Chen, X., Li, X., Shao, J., An, N., Tian, H., Wang, C., Han, T., Wang, L., Lu, B.: High-performance piezoelectric nanogenerators with imprinted P(VDF-TrFE)/$BaTiO_3$ nanocomposite micropillars for self-powered flexible sensors. Small **13**(23), 1604245 (2017)
3. Hu, C., Cheng, L., Wang, Z., Zheng, Y., Bai, S., Qin, Y.: A transparent antipeep piezoelectric nanogenerator to harvest tapping energy on screen. Small **12**(10), 1315–1321 (2016)
4. Yan, J., Jeong, Y.G.: High performance flexible piezoelectric nanogenerators based on $BaTiO_3$ nanofibers in different alignment modes. ACS Appl. Mater. Interfaces **8**(24), 15700–15709 (2016)
5. Anand, A., Bhatnagar, M.C.: Effect of sodium niobate ($NaNbO_3$) nanorods on β-phase enhancement in polyvinylidene fluoride (PVDF) polymer. Mater. Res. Express (2018)
6. Das, S., Biswal, A.K., Parida, K., Choudhary, R., Roy, A.: Electrical and mechanical behavior of PMN-PT/CNT based polymer composite film for energy harvesting. Appl. Surf. Sci. **428**, 356–363 (2018)
7. Kumar, C., Gaur, A., Tiwari, S., Biswas, A., Rai, S.K., Maiti, P.: Bio-waste polymer hybrid as induced piezoelectric material with high energy harvesting efficiency. Compos. Commun. **11**, 56–61 (2019)
8. Shin, S.-H., Kim, Y.-H., Lee, M.H., Jung, J.-Y., Nah, J.: Hemispherically aggregated $BaTiO_3$ nanoparticle composite thin film for high-performance flexible piezoelectric nanogenerator. ACS Nano **8**(3), 2766–2773 (2014)
9. Siddiqui, S., Kim, D.-I., Roh, E., Duy, L.T., Trung, T.Q., Nguyen, M.T., Lee, N.-E.: A durable and stable piezoelectric nanogenerator with nanocomposite nanofibers embedded in an elastomer under high loading for a self-powered sensor system. Nano Energy **30**, 434–442 (2016)

10. Gao, L., He, J., Hu, J., Li, Y.: Large enhancement in polarization response and energy storage properties of poly(vinylidene fluoride) by improving the interface effect in nanocomposites. J. Phys. Chem. C **118**(2), 831–838 (2014)
11. Mokhtari, F., Foroughi, J., Zheng, T., Cheng, Z., Spinks, G.M.: Triaxial braided piezo fiber energy harvesters for self-powered wearable technologies. J. Mater. Chem. A **7**(14), 8245–8257 (2019)
12. Hofmann, P., Walch, A., Dinkelmann, A., Selvarayan, S.K., Gresser, G.T.: Woven piezoelectric sensors as part of the textile reinforcement of fiber reinforced plastics. Compos. A **116**, 79–86 (2019)
13. Zhang, X., Li, B.-W., Dong, L., Liu, H., Chen, W., Shen, Y., Nan, C.-W.: Superior energy storage performances of polymer nanocomposites via modification of filler/polymer interfaces. Adv. Mater. Interfaces **5**(11), 1800096 (2018)
14. Mandal, D., Henkel, K., Schmeisser, D.: Comment on "preparation and characterization of silver–poly(vinylidene fluoride) nanocomposites: formation of piezoelectric polymorph of poly(vinylidene fluoride)". J. Phys. Chem. B **115**(35), 10567–10569 (2011)
15. Shi, K., Sun, B., Huang, X., Jiang, P.: Synergistic effect of graphene nanosheet and $BaTiO_3$ nanoparticles on performance enhancement of electrospun PVDF nanofiber mat for flexible piezoelectric nanogenerators. Nano Energy **52**, 153–162 (2018)
16. Hu, P., Yan, L., Zhao, C., Zhang, Y., Niu, J.: Double-layer structured PVDF nanocomposite film designed for flexible nanogenerator exhibiting enhanced piezoelectric output and mechanical property. Compos. Sci. Technol. **168**, 327–335 (2018)
17. Sri Abirami Saraswathi, M.S., Rana, D., Vijayakumar, P., Alwarappan, S., Nagendran, A.: Tailored PVDF nanocomposite membranes using exfoliated MoS_2 nanosheets for improved permeation and antifouling performance. New J. Chem. **41**(23), 14315–14324 (2017)
18. Kulkarni, S., Belavi, P., Khadke, U.: Synthesis, structural, characterization and dielectric spectroscopy of PVDF–$BaTiO_3$ polymer composite. In: AIP Conference Proceedings, p 090046. AIP Publishing (2018)
19. Lund, A., Rundqvist, K., Nilsson, E., Yu, L., Hagström, B., Müller, C.: Energy harvesting textiles for a rainy day: woven piezoelectrics based on melt-spun PVDF microfibres with a conducting core. NPJ Flex. Electron. **2**(1), 9 (2018)
20. Li, H., Liu, F., Fan, B., Ai, D., Peng, Z., Wang, Q.: Nanostructured ferroelectric-polymer composites for capacitive energy storage. Small Methods **2**(6), 1700399 (2018)
21. Kim, H., Lee, D., Kim, D., Kong, D., Choi, J., Lee, M., Murillo, G., Jung, J.: Dominant role of young's modulus for electric power generation in PVDF–$BaTiO_3$ composite-based piezoelectric nanogenerator. Nanomaterials **8**(10), 777 (2018)
22. Peng, R., Zhou, H., Wang, H., Mishnaevsky, L., Jr.: Modeling of nano-reinforced polymer composites: microstructure effect on Young's modulus. Comput. Mater. Sci. **60**, 19–31 (2012)
23. Yaqoob, U., Uddin, A.S.M.I., Chung, G.-S.: The effect of reduced graphene oxide on the dielectric and ferroelectric properties of PVDF–$BaTiO_3$ nanocomposites. RSC Adv. **6**(36), 30747–30754 (2016)
24. Nunes-Pereira, J., Sencadas, V., Correia, V., Cardoso, V.F., Han, W., Rocha, J.G., Lanceros-Méndez, S.: Energy harvesting performance of $BaTiO_3$/poly(vinylidene fluoride–trifluoroethylene) spin coated nanocomposites. Compos. B **72**, 130–136 (2015)
25. Alam, M.M., Ghosh, S.K., Sarkar, D., Sen, S., Mandal, D.: Improved dielectric constant and breakdown strength of γ-phase dominant super toughened polyvinylidene fluoride/TiO_2 nanocomposite film: an excellent material for energy storage applications and piezoelectric throughput. Nanotechnology **28**(1), 015503 (2016)
26. Kim, H., Renteria-Marquez, A., Islam, M.D., Chavez, L.A., Garcia Rosales, C.A., Ahsan, M.A., Tseng, T.-L.B., Love, N.D., Lin, Y.: Fabrication of bulk piezoelectric and dielectric $BaTiO_3$ ceramics using paste extrusion 3D printing technique. J. Am. Ceram. Soc. **102**(6), 3685–3694 (2019)
27. Martins, P., Lopes, A.C., Lanceros-Mendez, S.: Electroactive phases of poly(vinylidene fluoride): determination, processing and applications. Prog. Polym. Sci. **39**(4), 683–706 (2014)

28. Kar, E., Bose, N., Dutta, B., Banerjee, S., Mukherjee, N., Mukherjee, S.: 2D SnO_2 nanosheet/PVDF composite based flexible, self-cleaning piezoelectric energy harvester. Energy Convers. Manage. **184**, 600–608 (2019)
29. Thakur, P., Kool, A., Hoque, N.A., Bagchi, B., Khatun, F., Biswas, P., Brahma, D., Roy, S., Banerjee, S., Das, S.: Superior performances of in situ synthesized ZnO/PVDF thin film based self-poled piezoelectric nanogenerator and self-charged photo-power bank with high durability. Nano Energy **44**, 456–467 (2018)
30. Lu, X., Qu, H., Skorobogatiy, M.: Piezoelectric microstructured fibers via drawing of multimaterial preforms. In: Energy Harvesting and Storage: Materials, Devices, and Applications VIII, International Society for Optics and Photonics, p. 106630E
31. Song, Y., Shen, Y., Liu, H., Lin, Y., Li, M., Nan, C.-W.: Enhanced dielectric and ferroelectric properties induced by dopamine-modified $BaTiO_3$ nanofibers in flexible poly (vinylidene fluoride-trifluoroethylene) nanocomposites. J. Mater. Chem. A **22**(16), 8063–8068 (2012)
32. Sim, H.J., Choi, C., Lee, C.J., Kim, Y.T., Spinks, G.M., Lima, M.D., Baughman, R.H., Kim, S.J.: Flexible, stretchable and weavable piezoelectric fiber. Adv. Eng. Mater. **17**(9), 1270 (2015)
33. Lee, J.W., Cho, H.J., Chun, J., Kim, K.N., Kim, S., Ahn, C.W., Kim, I.W., Kim, J.-Y., Kim, S.-W., Yang, C., Baik, J.M.: Robust nanogenerators based on graft copolymers via control of dielectrics for remarkable output power enhancement. Sci. Adv. **3**(5), e1602902 (2017)
34. Sinha, T.K., Ghosh, S.K., Maiti, R., Jana, S., Adhikari, B., Mandal, D., Ray, S.K.: Graphene-silver-induced self-polarized PVDF-based flexible plasmonic nanogenerator toward the realization for new class of self powered optical sensor. ACS Appl. Mater. Interfaces **8**(24), 14986–14993 (2016)
35. Sun, B., Li, X., Zhao, R., Ji, H., Qiu, J., Zhang, N., He, D., Wang, C.: Electrospun poly(vinylidene fluoride)-zinc oxide hierarchical composite fiber membrane as piezoelectric acoustoelectric nanogenerator. J. Mater. Sci. Mater. Electron. **54**(3), 2754–2762
36. Mirvakili, S.M., Hunter, I.W.: Vertically aligned niobium nanowire arrays for fast-charging micro-supercapacitors. Adv. Mater. **29**(27), 1700671 (2017)
37. Karan, S.K., Bera, R., Paria, S., Das, A.K., Maiti, S., Maitra, A., Khatua, B.B.: An approach to design highly durable piezoelectric nanogenerator based on self-poled PVDF/AlO-rGO flexible nanocomposite with high power density and energy conversion efficiency. Adv. Energy Mater. **6**(20), 1601016 (2016)
38. Ding, R., Liu, H., Zhang, X., Xiao, J., Kishor, R., Sun, H., Zhu, B., Chen, G., Gao, F., Feng, X., Chen, J., Chen, X., Sun, X., Zheng, Y.: Flexible piezoelectric nanocomposite generators based on formamidinium lead halide perovskite nanoparticles. Adv. Funct. Mater. **26**(42), 7708–7716 (2016)
39. Yang, Y., Gao, Z.-S., Yang, M., Zheng, M.-S., Wang, D.-R., Zha, J.-W., Wen, Y.-Q., Dang, Z.-M.: Enhanced energy conversion efficiency in the surface modified $BaTiO_3$ nanoparticles/polyurethane nanocomposites for potential dielectric elastomer generators. Nano Energy **59**, 363–371 (2019)
40. Lin, Z.-H., Yang, Y., Wu, J.M., Liu, Y., Zhang, F., Wang, Z.L.: BaTiO3 Nanotubes-based flexible and transparent nanogenerators. J. Phys. Chem. Lett. **3**(23), 3599–3604 (2012)
41. Tamang, A., Ghosh, S.K., Garain, S., Alam, M.M., Haeberle, J., Henkel, K., Schmeisser, D., Mandal, D.: DNA-assisted β-phase nucleation and alignment of molecular dipoles in PVDF film: a realization of self-poled bioinspired flexible polymer nanogenerator for portable electronic devices. ACS Appl. Mater. Interfaces **7**(30), 16143–16147 (2015)
42. Lee, M., Chen, C.-Y., Wang, S., Cha, S.N., Park, Y.J., Kim, J.M., Chou, L.-J., Wang, Z.L.: A hybrid piezoelectric structure for wearable nanogenerators. Adv. Mater. **24**(13), 1759–1764 (2012)
43. Fu, J., Hou, Y., Gao, X., Zheng, M., Zhu, M.: Highly durable piezoelectric energy harvester based on a PVDF flexible nanocomposite filled with oriented $BaTi_2O_5$ nanorods with high power density. Nano Energy **52**, 391–401 (2018)
44. Xue, X., Wang, S., Guo, W., Zhang, Y., Wang, Z.L.: Hybridizing energy conversion and storage in a mechanical-to-electrochemical process for self-charging power cell. Nano Lett. **12**(9), 5048–5054 (2012)

Chapter 5
Highly Stretchable Self-Powered Wearable Electrical Energy Generator and Sensors

5.1 Introduction

Up to now, various piezoelectric materials have been discovered to enhance the energy harvesting performance through different polymer, ceramic, and conductive filler in a composite structure [1]. The disadvantage of these filler is that they are expensive, brittle, toxic, and non-environmentally friendly [2, 3]. Piezoelectric polymers have low dielectric constant and high breakdown strength [4]. In contrast, ceramic materials provide low breakdown strength and high dielectric constant. Hence, the combination of flexible polymer and ceramic fillers [5–7] make it feasible to improve the charge generation and storage capabilities of composite structure and have been considered for charge generation and storage applications [8]. During poling process, due to lower dielectric constant of piezoelectric polymers in compare with ceramics, the electric field has less effect on the piezoelectric ceramic particles, and they cannot be fully poled. To overcome this problem, the effective method is using conductive filler (graphene, carbon nanotubes, and metal particles) for creation a continuous electric flux path between the piezoelectric particles [9]. Graphene is an ideal nanofiller to enhance the electrical, mechanical, and thermal properties of polymers at very low loading contents [10]. The oxygen including in functional groups of GO and the PVDF polymer chains interact with each other and improve the dispersibility. From another side, the incorporation of GO into PVDF deteriorates the mechanical and electrical properties of the polymer composite. Therefore, the common method is to reduce GO to rGO, which still has graphene properties but enhanced its dispersion in the polymer [11]. The basal and edges planes of graphene nanosheets have oxygen functional groups, therefore, rGO can act as a negative triboelectric material. Moreover, rGO can limit electric charges in composite structure due to the difference between electrical conductivity of rGO and the matrix [12]. Reduced graphene oxide stabilizes polar phase in PVDF and leads the charges easily toward the electrodes and enhances the energy harvesting properties [13]. The electrostatic interactions between the positive and negative charge centres of metal ions and negative charge clouds in rGO leads to enhancement in the piezoelectric polar phases [11].

F. Mokhtari, *Self-Powered Smart Fabrics for Wearable Technologies*, Springer Theses,
https://doi.org/10.1007/978-3-031-06481-4_5

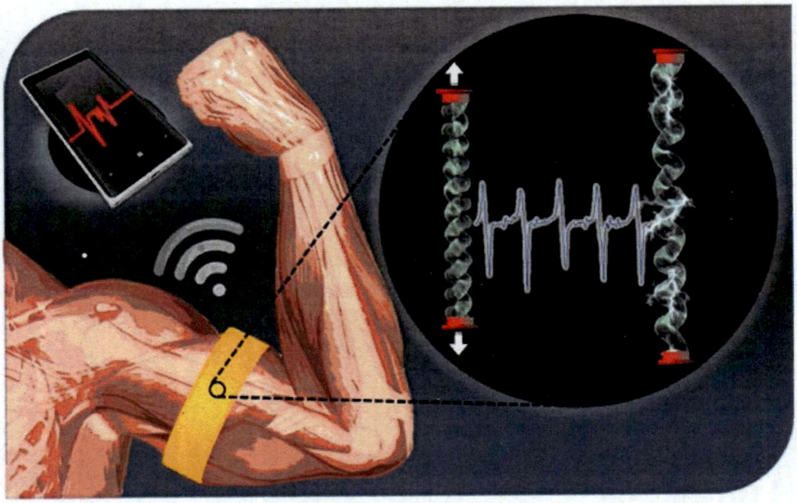

Fig. 5.1 Schematic illustration of the charge generation during longitudinal extension of PVDF/rGO/BT nanocomposite coil

The dielectric constant and ferroelectric behaviour of the nanocomposite confirm remarkable increase for the PVDF/rGO (0.1 wt%) in compare with pure PVDF [14]. The prepared composite film of PVDF and rGO (0.05 wt%) show more β-phase formation and three times higher output voltage than pure PVDF film [15]. In this chapter, the lightweight, flexible, lead free, cost effective and high performance PVDF based fiber were fabricated. PVDF, PVDF/BT, PVDF/rGO and PVDF/rGO/BT fibers were twisted until coils formed (Fig. 5.1), and then further twisted until the entire fiber length was coiled. The aim of this work is to characterise and investigate the effect of BT and rGO on the PVDF matrix. Also, the ferroelectric structural and piezoelectric properties of these nanocomposite fibers are described. The PVDF/rGO/BT nanocomposite coil present highest voltage output peak which is related to the β-phase dipoles alignment of the in PVDF owning to the existence of BT and rGO in addition to the natural piezoelectric properties of BT.

5.2 Experimental Details

5.2.1 Nanocomposite Fiber and Coil Preparation

Nanocomposite film of PVDF/BT and PVDF/rGO/BT were prepared as described in Sect. 2.2.4. Fiber spinning was performed using a twin-screw extruder as explained in Sect. 2.2.6. The composite chopped films (Fig. 5.2a) were heated overnight at a temperature of 70 °C then fed into the extruder. A single hole spinneret with a 2 mm diameter (Fig. 5.2b) was utilized to fabricate monofilament. The final fiber winding

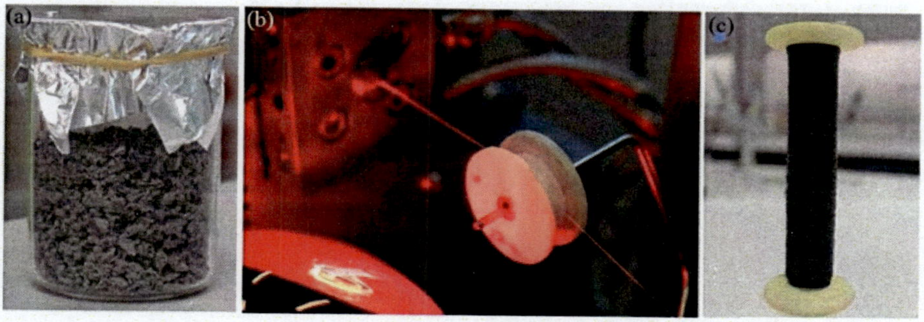

Fig. 5.2 Experimental procedure for the continuous production of nanocomposite fiber: **a** chopped PVDF/rGO film, **b** nanocomposite PVDF/rGO fiber coming out from the die to be collected, IR Lamp used for slow cooling as-spun fiber (80 °C) and **c** a spool of flexible PVDF/rGO nanocomposite fiber

speed of $v_w = 20$ m/min provides an inline cold drawing ratio $\lambda = v_w/v_t = 10$. The final diameter of the as-prepared PVDF fiber was ~170 μm (Fig. 5.2c). Cold drawing process was done on the fiber as described in Sect. 2.2.7. The fabrication process for the coil, woven and knitted structure was performed as explained in Sect. 2.2.9.

5.3 Results and Discussion

5.3.1 Morphology and Microstructures of the PVDF Nanocomposites

Figure 5.3a shows the dispersions of fillers in the transparent PVDF/DMF solution. Nanoparticles of BT ceramic were added into the PVDF matrix to improve the piezoelectric response. The optimized BT concentration has the value of 10 wt% based on our previous work [16] described in Chap. 4. Stable dispersions of rGO in DMF facilitated their mixing with the PVDF polymer, which can be readily dissolved in DMF as well. Figure 5.3b shows the images of the PVDF nanocomposites films. It can be seen that the introduction of either BT or rGO did not affect the formation of the PVDF films. The graphene nanosheets and the BT nanoparticles embedded homogeneously in the fibers and they are identifiable along the fiber axis (Fig. 5.3c). All prepared nanocomposite films and fibers are flexible. The flexibility of a PVDF/rGO nanocomposite film is shown in Fig. 5.3d. The nanocomposite fibers also have the ability for the knitting, weaving and braiding as reported in our previous research [17].

The surface morphological features of PVDF nanocomposite fibers were characterized by SEM. The surface morphology and cross-section (Fig. 5.4a–c) of nanocomposite fibers show that both melt-spun PVDF and hybrid PVDF/BT fibers are very smooth and without any observable porosity or voids. The homogeneous dispersion

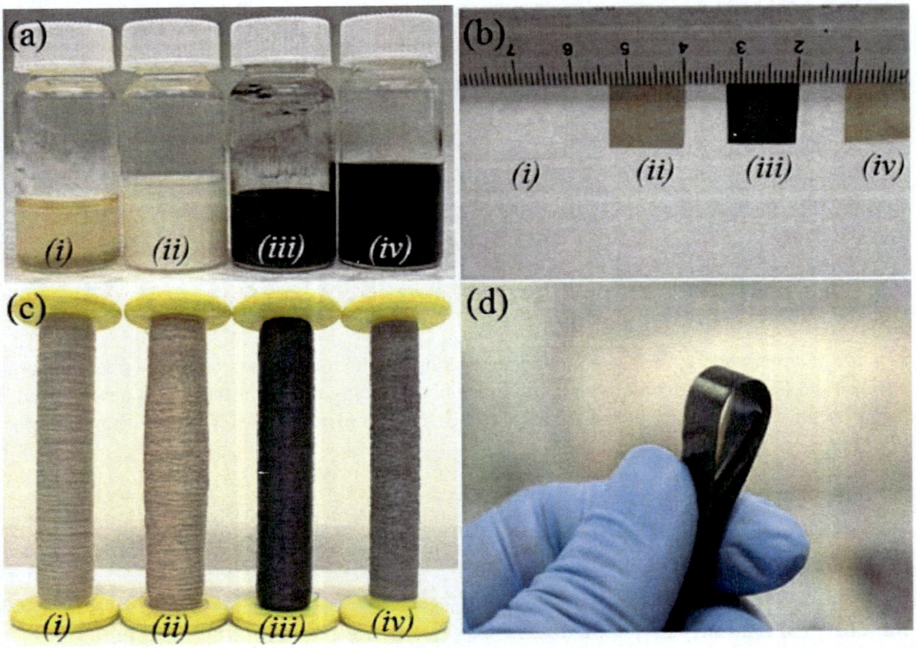

Fig. 5.3 Macro-pictures of **a** solution **b** film and **c** nanocomposite fiber of (i) pure PVDF, (ii) PVDF/BT, (iii) PVDF/rGO and (iv) PVDF/rGO/BT, **d** foldable flexible PVDF/rGO film

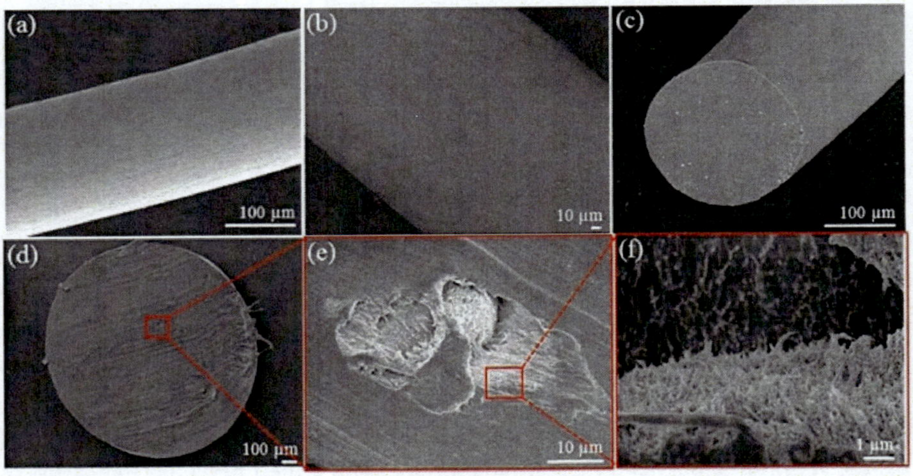

Fig. 5.4 SEM images of the as-spun nanocomposite fibers: **a** surface of PVDF fiber, **b** surface of PVDF/BT (10 wt%), **c** cross-section of PVDF/rGO/BT fiber, **d** cross-section of PVDF/rGO fiber at **e** low and **f** higher magnification

of the filler throughout the nanocomposite can be clearly observed in Fig. 5.4d-f from the cross section of the PVDF/rGO fiber. Figure 5.4d shows low contents of rGO sheet with the dimension of 200–400 nm. The graphene nanosheets can be seen in Fig. 5.4f. As described below, the addition of a low concentration (0.03 wt%) of rGO into PVDF enhanced the composite output performance. Using too high rGO concentration leads to agglomeration in the composites which create short circuits and cause charge leakage between electrodes [3]. The research show that only 0.1 wt% of rGO was sufficient to hamper α phase formation and promotea nearly complete β phase structure [18]. The energy harvesting performance and the high frequency capacitors improved for the rGO (0.3 wt%) and BT (35 wt%) of PVDF/rGO/BT nanocomposite film [19]. Hence, to evaluate the effect of rGO on the PVDF/BT nanocomposite and pure PVDF fiber, 0.5 wt% of the rGO selected for more consideration.

The piezoelectric performance of nanocomposites rely on the crystalline structure and the electroactive polar phase formation [20]. So, for clarification purposes of the polar crystalline phases present in the PVDF based nanocomposites, FTIR spectral analysis was performed, as presented in Fig. 5.5a. The α phase of PVDF has the peaks at 764 and 975 cm^{-1} [21]. These peaks nearly disappeared in the PVDF nanocomposite fibers with the presence of BT nanoparticles and graphene nanosheet which have an important role in the β phase formation [22]. The FTIR spectra of the melt spun fibers proves strong vibration peaks at 840 and 1275 cm^{-1}, which are related to the $–CH_2$-wagging vibration and $–CF_2$-symmetric stretching and the presences of these peaks demonstrate the β-phase formation in the composite [23, 24]. The reduced graphene causes formation of the piezoelectric phase in PVDF matrix as shown by peaks at 1173 and 1275 cm^{-1} as well as decreasing α-phase peak intensity. In fact, β-phase formation is mainly influenced by better interaction between the two components [25]. A high intensity peak at 1175 cm^{-1} in all nanocomposite fibers highlighted the high quality of the PVDF based fibers [19]. The intensity of vibration peaks is sensitive to the crystalline structure in polymer and the $F(\beta)$ of fiber structures was calculated from Eq. (3.1).

The variation of β phase contents for as-prepared PVDF and its nanocomposites fibers and coils are shown in Fig. 5.5b. More β phase formed in coil samples and this effect is likely due to more stretching and aligning of dipole along the fiber axis during coil fabrication. The highest β phase formation was present in the PVDF/rGO/BT nanocomposite coil with 58% improvement when compared with its untwisted fiber. The high fraction of β phase in PVDF based fibers are desirable due to their enhanced piezoelectric response. The β phase content here for PVDF/BT is different from Fig. 4.8b due to different fabrication process. However, each comparison between PVDF fiber and its composite done when they fabricated in the same condition. The FTIR and XRD technique had been used to explore effect of rGO and BT fillers on the crystalline structures of PVDF fiber. Figure 5.5c demonstrates the XRD patterns of the PVDF fiber and its combination with BT nanoparticles and rGO nanosheets. Pure PVDF has the stable α-phase (TGTG) peaks at $2\theta = 17.5°$ (100), 19.8° (110) and 26.5° (021). Other nanocomposite fibers have the same peak appearances owing to the existence of semi crystalline PVDF polymer [26].

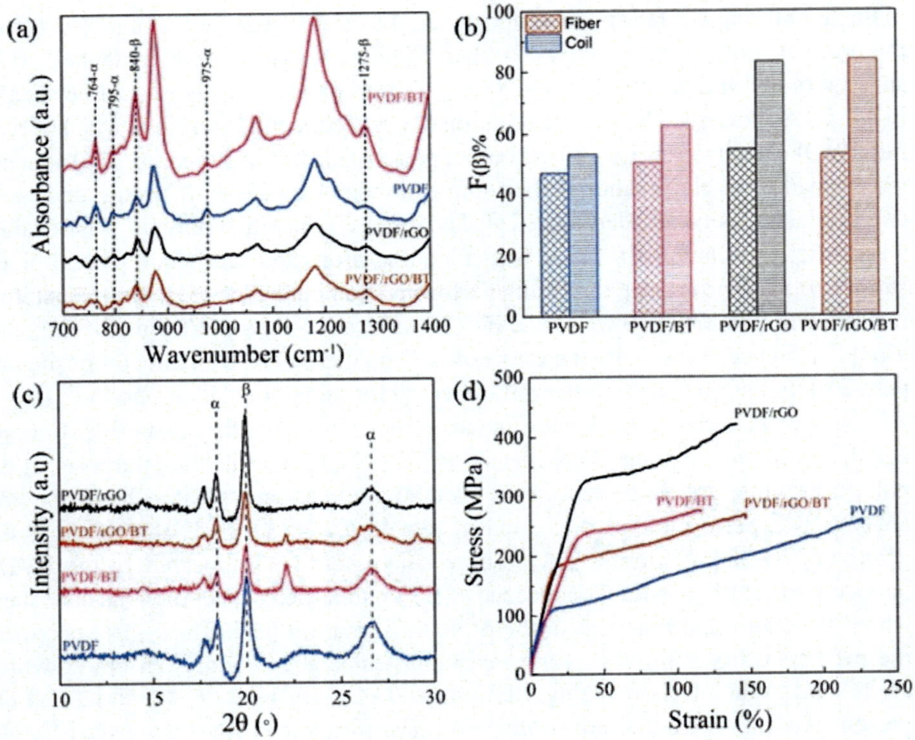

Fig. 5.5 Characterisation of the melt-spun PVDF, PVDF/BT, PVDF/rGO and PVDF/rGO/BT fibers: **a** FTIR spectra and **b** The difference in β phase contents of fibers and coils structure **c** X-ray diffraction patterns and **d** stress–strain curves obtained from tensile tests

The β phase (TTTT) appeared at 20° signifying the coexistence of α and β phase. The phase transformation from α to β in the nanocomposite fibers comes from the molecular chain stretching during the fabrication process. Another peak appears at 2θ–22° in BT nanocomposite fibers and is associated with the (001) and (100) reflections of tetragonal BT nanoparticle [22]. The layered graphite structure in rGO has diffraction peaks at 26.2° which is related to the (002) diffraction [27]. Also, an increase in the intensity of β phase peaks confirmed the good dispersion of the rGO in PVDF matrix [21]. The nucleation rate of the β phase in nanocomposites contained rGO increased gradually because of delocalization of π-electrons with remaining oxygen functionality in rGO which interact with the CH_2, CF_2 dipoles of the PVDF polymer chain [13]. Basically the interaction of OH groups with the fluorine (F) groups in the PVDF support them to be aligned on one side [28]. The deconvolution method was used to classify the existence of β and γ-phases individually. The incorporation of 0.5 wt% rGO increases the crystallinity (χ_{ct}) from ≈41% in PVDF (α ≈ 20.9%, γ ≈ 17.5%) to ≈43% (β ≈ 30%, γ ≈ 3%) in PVDF/rGO (0.5 wt%) composite. The as-prepared PVDF and its composite fibers have also been investigated to evaluate the effect of filler addition on the mechanical properties of the nanocomposite fibers. A comparison of the stress-strain curves for the

Table 5.1 Summary of mechanical properties of the as-prepared PVDF and PVDF/BT nanocomposite fibers

Sample	PVDF	PVDF/BT	PVDF/rGO	PVDF/rGO/BT
Young's modulus (MPa)	670	1593	1396	1127
Elongation at break (%)	226	117	141	146
Tensile strength (MPa)	258.5	276.7	420	268.5

various prepared fibers is given in Fig. 5.5d. While graphene is used as the filler in a composite structure, it is expected to have superior mechanical properties due to its high intrinsic strength. As can been seen from Table 5.1 and Fig. 5.6, the tensile strength and Young's modulus of the prepared composite fibers have significantly increased compared to PVDF fiber.

The Young's modulus and tensile strength of the as-prepared PVDF/rGO nanocomposite fiber is 208% and 163% higher than pure PVDF fiber, respectively. The tensile strength of the PVDF and PVDF/rGO nanocomposite fibers are 3 and 6 times higher than fibers reported recently [29]. The elongation at break for the as-prepared PVDF and PVDF/rGO nanocomposite fibers are 226% and 141%, respectively. These values are 15 and 57 times higher than previously reported for the PVDF and PVDF/rGO nanocomposite film [30]. Adding 10 wt% BT produces fibers with an elongation at break of 120%. This value is 3 times higher than previously reported for melt spun PVDF fiber [31]. The stress–strain curve for the PVDF/rGO fibers show fast-rising trend in stress within a strain of 30% due to stress concentration caused by the rigid rGO lamella networks. When the strain is over 30% a moderate increase was observed which indicates that the rigid rGO lamella networks are gradually broken up [32]. The enhancement in the β crystallites of the composite afford advancement in the mechanical properties (confirmed by FTIR measurements) [33].

One of the reasons to have better mechanical properties in composite structures is a good dispersion of particles with strong interfacial interaction between the filler particles and the polymer matrix. The BT nanoparticles have desirable electrostatic

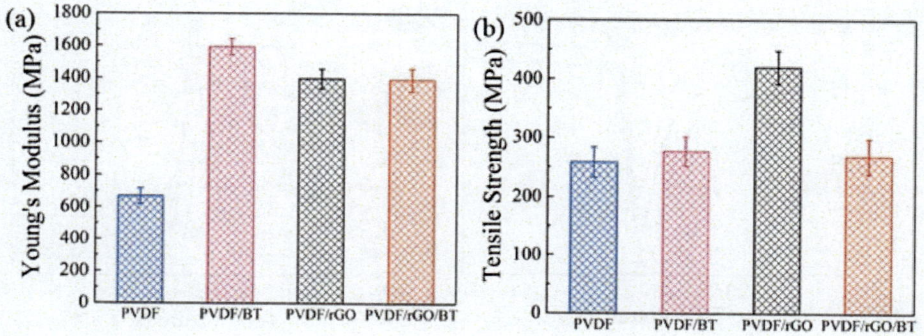

Fig. 5.6 Mechanical characterization of PVDF, PVDF/BT, PVDF/rGO and PVDF/rGO/BT nanocomposite fibers: **a** Young's modulus and **b** tensile strength

interaction/hydrogen bonding with PVDF chains due to their high dipole attraction. Therefore, BT nanoparticles acts as a 'bridge' to connected PVDF chains and rGO particles [34]. However, the intimate interaction between BT nanoparticles and rGO substrates can also limit the dispersion of BT nanoparticles in rGO sheets and cause aggregation formation. Addition of rGO/BT fillers into the PVDF matrix can lead to a crowding effect and these unexfoliated aggregates create stress concentration points during the propagation of the cracks and make lower tensile strength [35].

The effect of the nanofiller addition on the crystallization and melting temperatures of the nanocomposites were determined by differential scanning calorimetry (DSC) analysis. The second heating and cooling characteristics of the samples at a scanning rate of 5 °C/min were monitored and the curves obtained are depicted in Fig. 5.7. The nanoparticles effects on the thermal properties of PVDF/BT is clearly seen from the peak shifts. The melting temperature (T_{m}) and crystallization temperature (T_{c}) slightly increase in the nanocomposite by adding the filler (Table 5.2). This increment is a result of the homogeneous dispersion of filler throughout the matrix, which has the role of nucleating agent and prevents segmental movement of the polymer chain [13]. The crystallinity percentage (χ_{c}) was determined from the following Eq. (5.1):

$$X_c = \frac{\Delta H_m}{(1-\varphi)\Delta H_m^0} \times 100 \tag{5.1}$$

where, ΔH_{m} is the nanocomposite melting enthalpy, ΔH_m^0 is the melting enthalpy of the 100% crystalline PVDF (103.4 Jg^{-1}) and ϕ is the weight percentage of the nanocomposite filler [18].

The pure PVDF showed a melting peak at 172.4 °C, which was slightly lower than that for PVDF/rGO nanocomposites at 172.6 °C. The PVDF β phase has higher melting temperature than α phase due to has *trans* conformation which let the chains to be packed more compactly in crystal structure [18, 36]. As a result, the predominant phases for the pure PVDF and PVDF/rGO nanocomposites were mainly α phase and

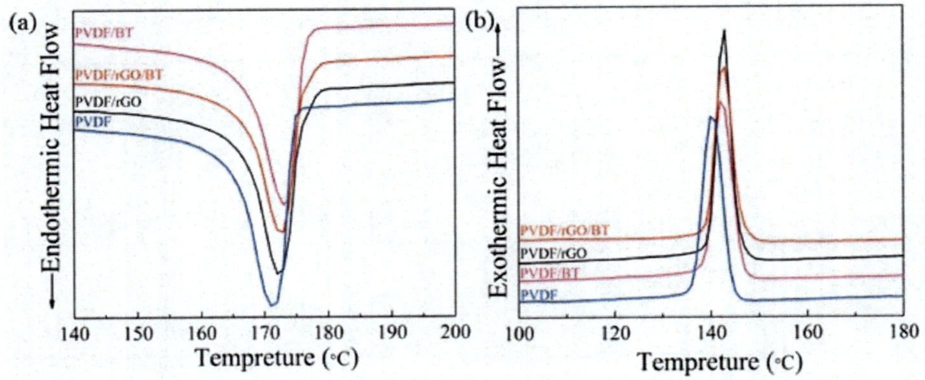

Fig. 5.7 DSC second melting (**a**) and cooling (**b**) curve of PVDF, PVDF/BT, PVDF/rGO and PVDF/rGO/BT nanocomposite fibers

Table 5.2 DSC results for PVDF and its nanocomposites fibers

Samples	T_m (°C)	T_c (°C)	X_c (%)
PVDF	172.4	140.1	41
PVDF/BT	173.3	142.4	42
PVDF/rGO	172.6	143.3	43
PVDF/rGO/BT	172.5	143	43

β phase subsequently which was confirmed by FTIR and XRD results. The cooling DSC curve indicated that crystallization happened at a higher temperature (Tc) for nanocomposites fibers in compare with pure PVDF fiber.

The crystallization temperature increased for the composites included rGO due to very large surface area of rGO which adsorbs PVDF chains and cause easier nucleation [18]. The PVDF/rGO/BT and PVDF/rGO nanocomposites exhibit similar crystallization behaviours. The difference between them is that the PVDF/rGO nanocomposite has higher Tc (143.3 °C) compared with the PVDF/rGO/BT nanocomposite (143 °C). The possible reason can be changing in the dispersion quality of the BT particles after adding rGO nanosheet. Many of the BT particles are deposited on the rGO sheet and consequently reduced PVDF crystallites growth [37]. The crystallinity temperature variation for nanocomposites depend on the dimensions, filler concentration and the interfacial interactions [13].

Thermal stability of the composite fiber was characterized by the thermogravimetric analysis (TGA) which was performed in temperature range of 30–900 °C and is presented in Fig. 5.8. For temperatures lower than 400 °C, the thermal stability of the nanocomposite contained rGO increased. The increased thermal stability of the rGO was caused by the high thermal conductivity of rGO, which facilitates dissipation of the thermal energy very quickly [38]. For rGO, the slight mass loss at the range of 200–400 °C is attributed to the removal of residual oxygen containing

Fig. 5.8 Thermographs of PVDF, PVDF/BT, PVDF/rGO and PVDF/rGO/BT nanocomposite fibers

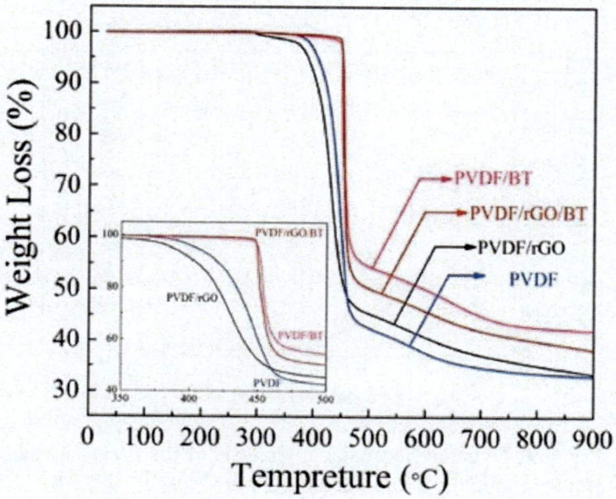

groups on rGO surfaces [39]. The higher thermal conductivity of the rGO helps to transfer heat from the GO layers to the PVDF matrix [40]. Temperature related to 5% degradation is considered as the onset of thermal degradation ($T_{5\%}$). Both the $T_{5\%}$ and the temperature corresponding to 50% weight loss ($T_{50\%}$) increase by up to 43 and 44 °C, respectively, in the composite of PVDF/rGO/BT in comparison with pure PVDF. This enhancement in thermal stability comes from a variety of physico-chemical interactions among composite components [41]. Pure PVDF fiber begins to decompose at about 370 °C, but the PVDF/BT and PVDF/rGO/BT nanocomposite fibers appear to degrade at 450 °C. The strong interaction between nanoparticles and the PVDF matrix was indicated by the XRD patterns and FTIR spectra. Furthermore, the residual weight at 900 °C of PVDF nanocomposites increases clearly in comparison with the pure PVDF according to their concentration. The residual mass for pure PVDF is 33.15% and is 33.59% for the PVDF/rGO and this weight difference confirms the addition of 0.5% of rGO in the composite structure. Also, the residual mass for PVDF/BT nanocomposite is 42.32% which confirms a 10% addition of BT in the composite.

In Fig. 5.9, the storage modulus and the loss factor at frequency 1 Hz are shown as a function of temperature from −50 to 150 °C at a heating rate of 3 °C/min for all PVDF nanocomposite fibers. The storage modulus depicts material ability for energy storage when an oscillatory force is applied. The loss factor (tan δ) depends on material viscosity and the tan δ peak arises from cooperative motion of different segments of the polymer [42].

From Fig. 5.9a it is clear that by increasing temperature the storage modulus of PVDF nanocomposites gradually decreases in a nonlinear way that signifies a progressive loss of the nanocomposites resistance to elastic deformation [18]. The storage modulus for pristine PVDF composites was found around 4 GPa at −50 °C and decreased with temperature over the investigated temperature range (−50 to 150 °C). All other composites exhibited temperature dependences, which are similar to the neat PVDF. When the temperature is below 25 °C all nanocomposites have

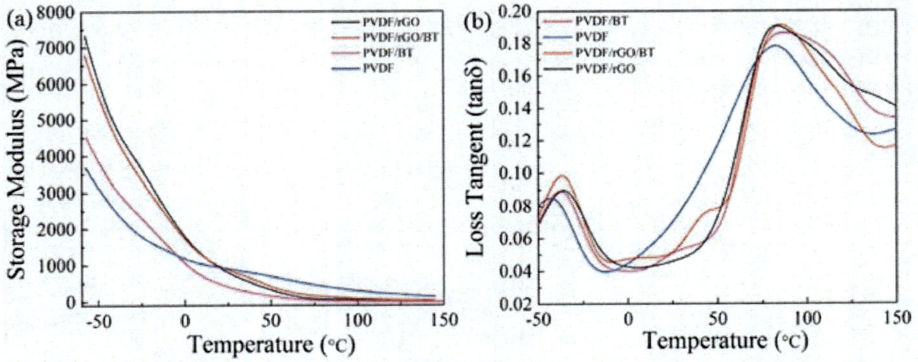

Fig. 5.9 Dynamic mechanical analysis of the PVDF nanocomposite fibers at frequency 1 Hz: **a** storage modulus, **b** loss tangent (tan δ)

higher storage moduli than the pure PVDF with the enhancement coming from rGO reinforcing effect. Sheets of rGO have high aspect ratio with strong interfacial adhesion for the integration in the PVDF matrix [43]. Although for temperature above 25 °C due to the movement, dispersion and directional distribution of composite particles in fibers during DMA test the PVDF fibers shows higher storage modulus. Figure 5.9b shows two relaxation peaks at low and high temperature for PVDF based fibers. PVDF exhibits peaks at −44.9, and 85.5 °C corresponding to T_g and T_r. The T_g is the glass transition temperature resulting from the start of segmental motion of PVDF amorphous portion. The T_g increases with increasing filler loading because of strong cohesive force between the filler which make the polymer more compact and decrease the free space for polymer segmental motion. The T_r indicates relaxation of PVDF chains at the crystalline region where the segments are in limited state [44]. The position of the T_r peak for the pure PVDF and the PVDF nanocomposites is different. The increase of difference in transition temperatures for relaxation in crystalline region (T_r) could be ascribed to the PVDF polymorphic structure variation from the α to β phase. The β phase has more dense volume (density = 1.99 g/cm^3) compared to the α phase (density = 1.92 g/cm^3) and amorphous phase (1.78 g/cm^3), hence it needs more thermal energy for the relaxation. A sharp peak of relaxation temperature happen for PVDF nanocomposites when the β phase conversion is complete [45]. A shoulder peak appears at around 50 °C for PVDF/rGO/BT which confirm the motion of molecular chain segments between the composites filler of BT an rGO resulting in thermal stimulated charge carriers.

5.3.2 *Ferroelectric Properties of Nanocomposite Films*

The polarization—electric field loop measurements has been carried out at room temperature to discover ferroelectric behaviours in the PVDF and nanocomposite films under an electric field of 300 kV/cm. The P–E loops for pure PVDF, PVDF/BT, PVDF/rGO and PVDF/rGO/BT composite films are presented in Fig. 5.10. The electrode area was 0.1962 cm^2, and the film thickness was 150 μm. The formation of the loops come from the presence of phase separation between voltage and charge. By increasing the applied electric field, the polarization gradually increases as wellbecause of the alignment of molecular dipoles in one direction [11]. The PVDF/rGO/BT nanocomposite film shows a maximum polarization value of 0.71 μC cm^{-2} when the applied electric field is 300 kV cm^{-1} (Fig. 5.10a). The PVDF/rGO/BT nanocomposite film shows the remnant polarization of 0.12 μC cm^{-2}, while it is 0.032 μC cm^{-2} for pure PVDF. The variation in remnant polarization could refer to molecular dipoles charge accumulation. This charge accumulation comes from the interactions between PVDF chains and the oxygen functional groups of rGO–BT nanocomposite. Therefore, the *P–E* hysteresis loops of the PVDF/rGO/BT nanocomposite film were assessed from 50 to 350 kV/cm until breakdown (Fig. 5.10b). The remanent polarization (Pr) of the nanocomposite films increased from 0.088 for PVDF

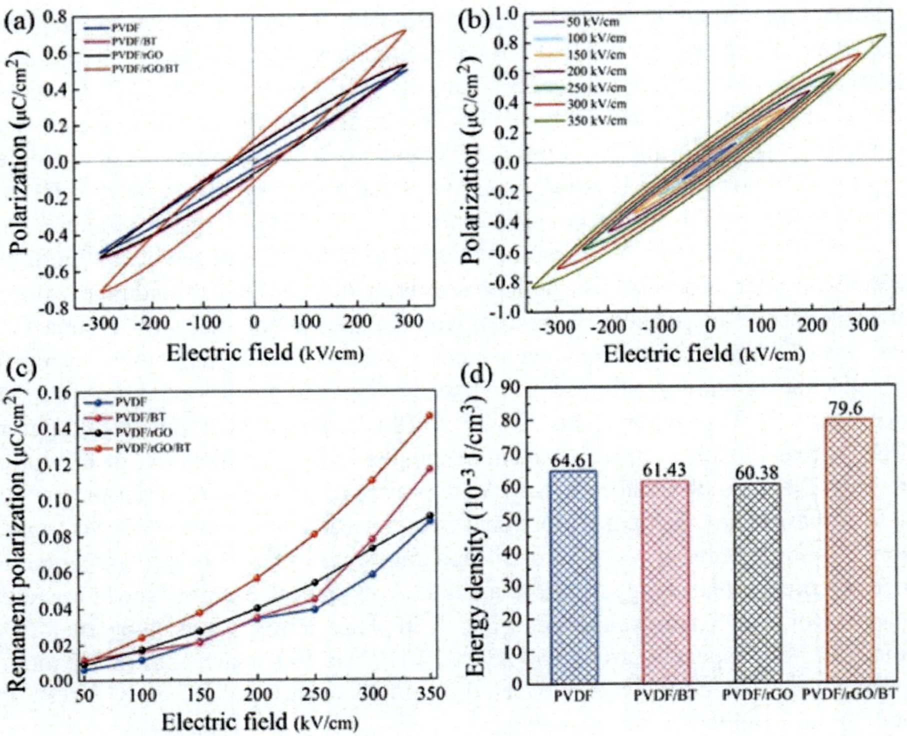

Fig. 5.10 Polarization—electric field (P–E) hysteresis loops of: **a** the PVDF, PVDF/rGO, PVDF/BT and PVDF/rGO/BT nanocomposite films, **b** the PVDF/rGO/BT nanocomposite film measured from 50 to 350 kV/cm, **c** remanent polarization at different electric fields and **d** energy density for different nanocomposite films

to 0.15 μC/cm^2 for PVDF/rGO/BT at 350 kV/cm and 1 Hz, which indicated ferroelectricity enhancement in the nanocomposites (Fig. 5.10c). Furthermore, remnant polarization is just in accordance with the piezoelectric response of material [13]. Whereas, the PVDF/rGO/BT nanocomposites has higher remnant polarization than PVDF, it reveal higher output voltages. Therefore, the nanocomposites are excellent for piezoelectric energy harvesting applications and also provide a high energy density capability. Since piezo ceramic BT can be polarized at higher fields, at the electric field above 300 kV/com shows higher remanet polarization than PVDF/rGO and PVDF. This trend observed for the PVDF/rGO/BT composite from beginning due to presence of conductive filler rGO which helps to polarized piezoelectric component better. The real energy density of PVDF nanocomposites to determine the energy harvesting strength could be calculated from the *P–E* loop by the following Eq. (5.2):

$$U_e = \int E dP \tag{5.2}$$

where E is the electric field and P refers to the polarization of the samples [46]. In fact, the energy density relies on the dielectric constant, dielectric breakdown strength, polarization and the applied electric field [47]. The energy density greatly increased up to 79.6 mJ/cm^3 at 300 kV/cm with nanocomposite containing PVDF/rGO/BT, which is 23% higher than pure PVDF (64.61 mJ/cm^3). The strong interactions between the rGO/BT and PVDF matrix could be the main reason for the polarization properties. The energy density graph in Fig. 5.10d shows that the charges at the interface for the PVDF/rGO/BT nanocomposite is much higher than the PVDF/BT due to the conductive nature of the rGO sheet. In fact presence of rGO in the structure helps to better charge transferring to the both BT and PVDF for polarization which leads to have more energy density between PVDF/rGO/BT in compare with PVDF/BT and PVDF/rGO. Nano particles of BT need higher electric field for polarization and need a conductive filler inside the composite structure due to this reason the energy density of PVDF uniform structure is a little bit more that PVDF/BT and PVDF/rGO.

5.3.3 Dielectric and Conductivity Properties of Nanocomposite Fibers

The dielectric and conductivity behaviour of the nanocomposite fibers at different frequencies at room temperature was evaluated and the results are shown in Figs. 5.11 and 5.12. The dielectric constant of the nancomposite fibers increased with adding BT, since BT possess a higher dielectric constant compared with pure PVDF polymer (Fig. 5.11a) [48]. The rGO introduces an active dielectric interface between the PVDF and the BT and induces orientational polarization in PVDF/rGO/BT nanocomposite.

It is clear that adding rGO induces a higher dielectric constant, dielectric loss and conductivity [49]. After adding rGO, many micro-capacitors formed inside the nanocomposites which leads to increase in dielectric constant [19]. The theory of percolation described the sharp rise in dielectric constant of the PVDF/rGO (0.5 wt%) through the following Eq. (5.3):

$$\varepsilon = \varepsilon_m (f_c - f)^{-s} \quad for\ f < f_c \tag{5.3}$$

where the dielectric constant of the composites (ε) and polymer (ε_m) are used in Eq. (5.4). The f and fc represent the concentration and the percolation threshold concentration of the filler in the matrix, respectively. The s is a constant which depend on materials properties [50].

When f gets close to fc in Eq. (5.3), the dielectric constant being extraordinarily high. Moreover, based on microcapacitor model, two neighbouring rGO fillers act as electrodes on both side of thin layer of PVDF [49]. Each microcapacitor has a large capacitance due to increase in dielectric constant. According to the micro-capacitor model, the maximum dielectric value comes from dipole alignment in one direction. It is also shown by Huang et al. [18] that dielectric loss and dielectric constant

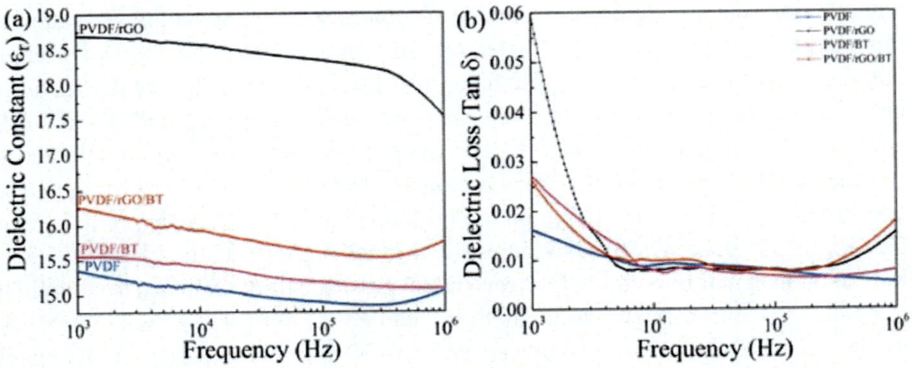

Fig. 5.11 The frequency dependence of **a** the dielectric constant and **b** the loss tangent of the PVDF, PVDF/rGO, PVDF/BT and PVDF/rGO/BT nanocomposite fibers at room temperature

slightly increase at lower rGO concentration, which is a common feature of insulating materials. While the rGO's loading value was 0.5 wt% and above, the dielectric loss and dielectric constant showed a significant increase [18]. At low frequency, the dielectric constant has a higher value in comparison with high frequency. At low frequency, the high dielectric constant refer to Maxwell–Wagnar–Sillars (MWS) interfacial polarization as a result of the heterogeneous conductivity of PVDF and rGO [51]. The growth of conductive grain boundary leads to dielectric constant decreasing at high frequency. Besides, dielectric constant and loss have lower values at a high frequency because of polarization effects and dipoles' inability to achieve the equilibrium state during electric field variation [52]. Figure 5.11b presents the dielectric loss (tan δ) variations versus frequency at room temperatures. Along with incorporation of rGO nanosheet, both the dielectric constant and the dielectric loss tends to increase [1]. In the dielectric material, the dielectric loss is defined as the energy dissipation through dipolar loss, conduction loss and interfacial polarisation [53]. By increasing frequency, dielectric loss decreases due to polarization effect (MWS effect). Additional reason could be the outset of the field variation tracking by the dipoles at higher frequencies [54]. In general, the smaller dielectric loss and the higher dielectric constants are the regular behaviour for the PVDF/rGO nanocomposites. Since rGO affords a greater number of conductive pathways into the PVDF matrix, it can cause an increase in dielectric loss due to the leakage current [46].

The conductivity of all nanocomposite fibers is shown in Fig. 5.12. The conductivity roughly shows a linear increase with increasing frequency. It is obvious that the nanocomposites filled with high conductive material (rGO) have the highest conductivity among other nanocomposites. The homogeneous rGO dispersion leads to have a continuous conducting network path across the nanocomposite. Therefore, the nanocomposite exhibits a higher surface conductivity and sensitivity even at low filler concentration [13]. At high frequency, the conductivity increase because of charges

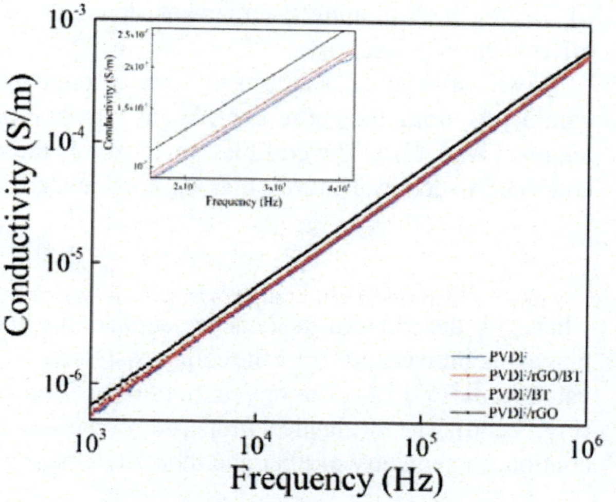

Fig. 5.12 Graph of conductivity as a function of frequency for PVDF nanocomposite fibers at room temperature

excitation and their transportation inside the composite [55]. The PVDF/rGO sample confirms the maximum conductivity of 6.11×10^{-4} (S m^{-1}) at 1 MHz frequency.

5.3.4 Performance of Nanocomposite Coil Structures

Figure 5.13 shows optical microscope images of coil samples fabricated from pure PVDF. The coil has a diameter (*D*) of ~270 μm and its filament is 160 μm (Fig. 5.13a,

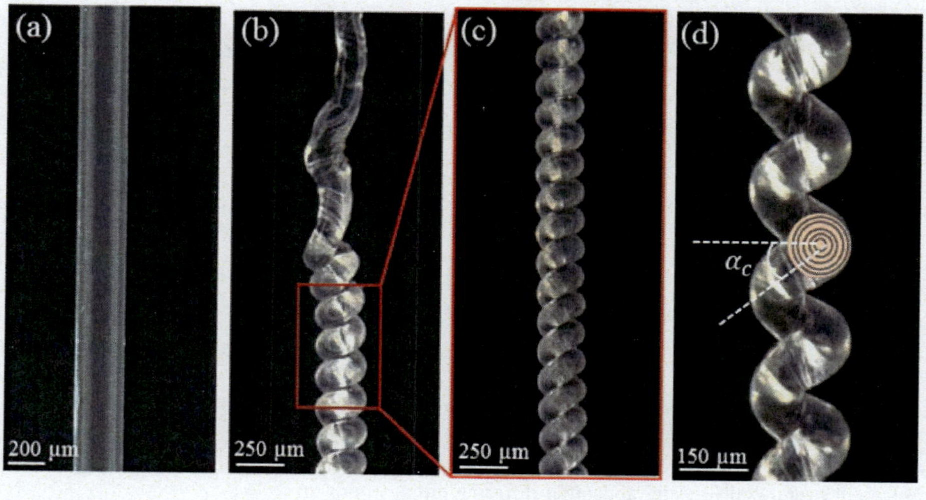

Fig. 5.13 Optical micrographs of **a** PVDF meltspun fiber, **b** coil formation process by inducing twist, **c** the magnified coil section, and **d** coil bias angle from close view of coil section

b). During meltspinning the filaments tends to be elongated and oriented in the microfiber direction.

A twist of about 12,000 turns m^{-1} was inserted into the filaments until coil formed completely along the entire fiber length. The coil has a uniform structure along its length (Fig. 5.13c). The coil bias angle (α_c) is the angle between the fiber and the coil's cross-section. The coils bias angle calculated from the following Eq. (5.4) [56]:

$$\cos(\alpha_c) = \pi ND/l \tag{5.4}$$

where N is the coil turns and l is the length of the fiber which made the coil. The coil bias angle showing 30° from the coil's cross-section, which agrees with the calculated value from Eq. (5.4). The optical microscope was used to observed coil bias angle (Fig. 5.13d). The mechanical properties of nanocomposite samples improved due to continuous geometry of fiber and their stretchability.

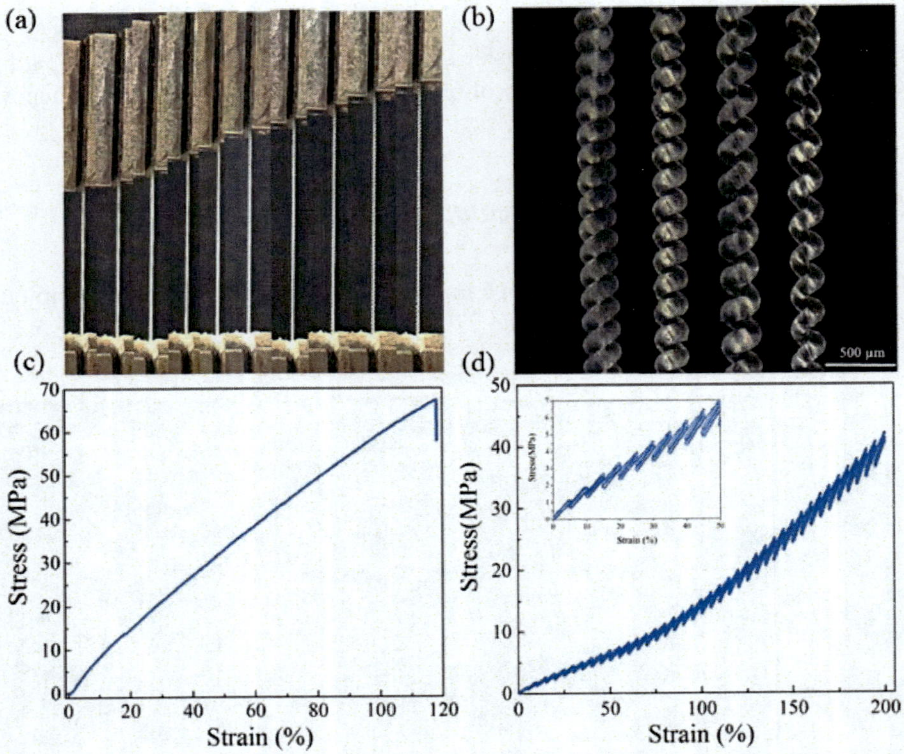

Fig. 5.14 **a** Digital photograph of the stretching procedure of a PVDF coil showing the coil expanding under tension up to ~120% strain. **b** Optical microscope image of a coil under different strain. **c** Stress-strain curve for pure PVDF coil. **d** Stress-strain curves of the loading and unloading filament over a 200% strain range after differing initial strains

The mechanical properties of fabricated coils were explored through the uniaxial tensile test of samples (Fig. 5.14a, b). The tensile extension in the longitudinal direction for the coil structure shown in Fig. 5.14c. To assess coil performance, the samples were vertically stretched several times. Figure 5.14d shows stress-strain curves found on loading and unloading filament over a 200% strain. The coil has significantly lower stiffness (200% strain at 40 MPa) compared with its fiber (50% strain at 100 MPa). Therefore, coil structures are more desirable for stretchy fabrics to monitor movement of robots, human and prosthetics without limitation or everywhere that a continuous stretching is considered. It is also possible to use more than one fiber to make a coil in order to get a higher piezoelectric response and higher tensile strength. The diameter of the coil structure depended on the number of fibers used in the supply coil. The coil diameter for single and double 2-ply fibers are ~270 and ~500 μm (Fig. 5.15a). In spite of a large amount of twisting, the fibers had no signs of failure and the fiber seems to be continuous. Although coils made from 2-ply fibers were highly stretchable, it was noted that deformation occurred by micro-buckle unfolding rather than coil opening and this process lead to a greater variation in voltage output in the 2-ply coils in comparison with the single fiber coil especially as higher strain (Fig. 5.15b). For all PVDF nanocomposites, the output voltage increased for 2-ply fiber coil in comparison with the single fiber coil. The higher amount of output voltage related to the PVDF/rGO double fiber coil with ~90% increment. This higher amount of value can be described by the higher conductivity of PVDF/rGO nanocomposite which facilitate charge transfer from coil surface to both electrodes. It is shown that coils harvesters provided arbitrarily high voltages if multiple harvesters were combined in series [57]. All fabricated coils were exposed to stretching force and the effect of different fillers studied. Figure 5.16a, b shows the open circuit output voltage and short circuit current for all PVDF nanocomposite. Among all the samples, the PVDF/rGO/BT nanocomposite showed a maximum voltage output of 1240 mV at an

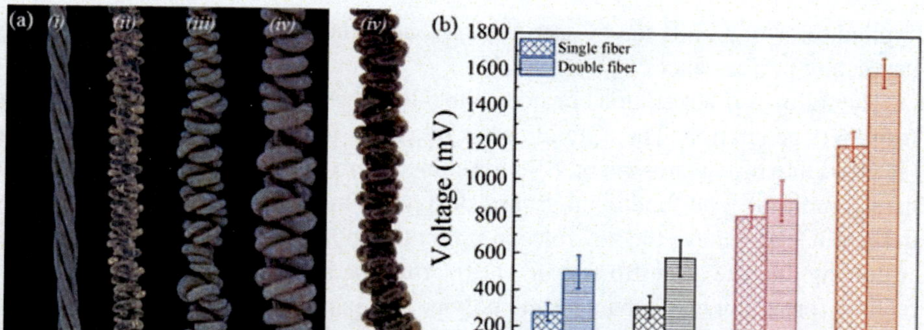

Fig. 5.15 **a** Optical microscope image of double fiber coil: (i) highly twisted fibers (ii) PVDF, (iii) PVDF/BT, (iv) PVDF/rGO/BT, (v) PVDF/rGO, **b** output voltage for single and double fiber coil structure after 100% strain at frequency of 1 Hz

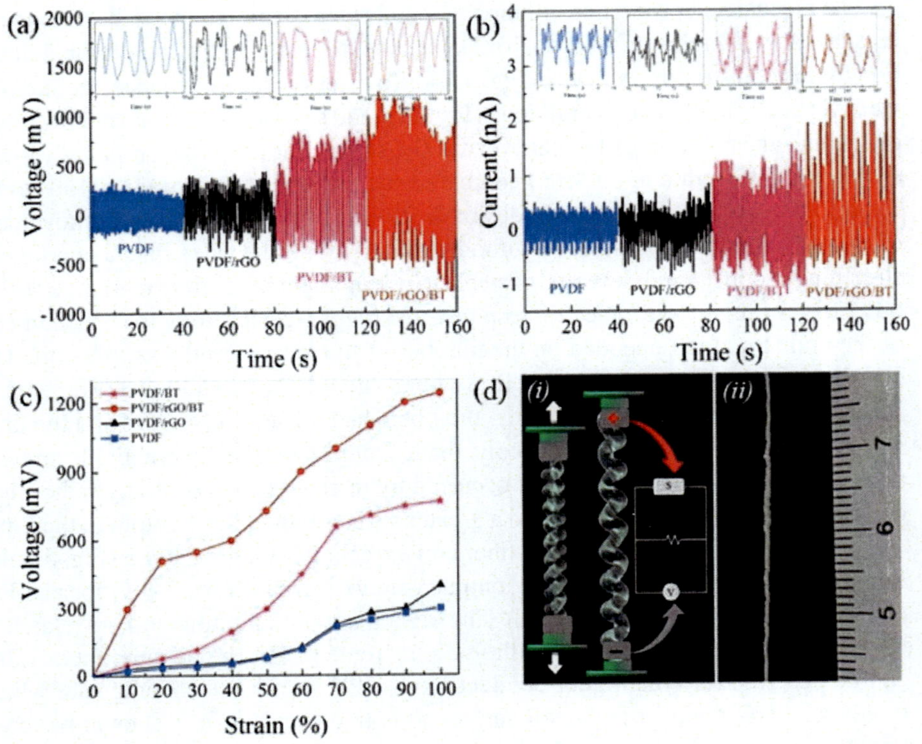

Fig. 5.16 Electric output of the PVDF and PVDF nanocomposites for 1 cm of coil structure during axial extension: **a** the signals of open-circuit voltage and **b** short-circuit current at frequency of 1 Hz at strain 100%, **c** output voltage for different percentage of strain and **d** coil structure: (i) Schematic illustration of the coil generator with aluminium foil on both sides as electrodes (ii) image of a coil with several centimetres long

applied force of 1 N. It can be seen that rGO contribution enhances the coils energy harvesting performance.

The following description considers the rGO function in nanocomposite structure for power generation. The rGO filler has two roles in the nanocomposite structure: (1) providing dipole alignment, (2) prompting micro-capacitor formation within the nanocomposite. The functional groups of oxygen including carboxyl and carbonyl in the rGO plane have the key roles in promoting polymer chain alignment through collecting fluorine atoms to one side [19]. After loading rGO to the composites, the β phase formation was enhanced and polymer chain alignment happened which was indicated according to the XRD results. The highly dispersed rGO sheets increases the availability of many free charges on their surface. In the PVDF/BT nanocomposite structure, both PVDF and BT dipoles accumulated free charges of rGO, and new dipoles have been generated in those specific regions. In comparison with other conductive fillers, rGO has higher surface area. Hence, more dipole would be formed and as a result the storage capability and charge generation will be enhanced [1]. The pure PVDF coil generated a maximum voltage output of 300 mV while

PVDF/rGO and PVDF/BT samples showed 400 mV and 700 mV maximum output voltages, respectively. The output voltage for PVDF/BT nanocomposite is higher than PVDF/rGO, this may due to number of rGO sheets and many free charge of graphene sheets which did not find dipoles from composite and these free charges stay inside the composite and negatively affected coil performance [19]. The results show that the piezoelectric output of coils were more sensitive to the strain percentage than the loading rate. In frequencies higher than 1 Hz, the short equilibrium time does not let the coil to reach stable surface charge transference between electrodes and this negatively affected the generated signal. As illustrated in Fig. 5.16c, there exists an output voltage of 300 mV when a strain of 10% was applied on PVDF/rGO/BT nanocomposite coil. When the strain increased to 100%, the output voltage reaches 1.2 V which is ~4 time higher than in the PVDF coil. Additionally, the nanocomposite coils has high stretchability up to 100%, which is significantly higher than recent report for piezoelectric fibers (<50%) [58]. The relation between strain and output voltage is relatively linear for all nanocomposites. The mechanism to show voltage generation by coil structure is illustrated in Fig. 5.16d. The power generated by the nanocomposites coils was calculated through Eq. (4.3) [58]. The maximum power output for 10 mm length PVDF/rGO/BT nanocomposite coil reached 3.6 μW, 0.42 $\mu W\ cm^{-3}$ or 3 $W\ kg^{-1}$ based on considering the diameter, length and mass of the coil. The energy conversion efficiency from mechanical energy into electrical energy was 22.5%, which was extracted from Eq. (4.6). This conversion efficiency for PVDF/rGO/BT nanocomposite coil is ~2.5 higher than latest reported harvesting electrical energy from coil structure of polyurethane microfibers [59].

5.4 Application Performance of Nanocomposite Fibers

The potential applications of the PVDF/rGO fiber have been explored in different application for movement sensor as can be seen from Figs. 5.17, 5.18, 5.19 and 5.20. It is possible to use other nanocomposite fibers of PVDF/BT and PVDF/rGO/BT in the same applications regards to their properties mentioned in previous sections. Although some applications for PVDF/BT nanocomposite fibers mentioned in Sect. 4.3. The PVDF/rGO fiber fabricated in the plain weave structure (4 × 10 cm with total weigh of 3 g) with silver coated nylon as electrodes and polyester fibers to separate each unit of generators is shown in Fig. 5.17a, b. The woven structure has four active units (4 × 1 cm, 0.25 g) and where the electrodes can be connected in series or in parallel connection. The woven generator was poled for serial and parallel connections. It was also found that the output voltage and current of the PVDF nanogenerator could be enhanced by serial and parallel connections, respectively [60].

For both series and parallel connections, the output peak voltage increases with increasing the number of active units. Woven piezoelectric fibers units in a series connection exhibits a higher output peak voltage and power than that of a parallel connection (Fig. 5.17c). For the series connection, the maximum output peak voltage for four active unit is 1000 mV under periodic pressure and relaxation.

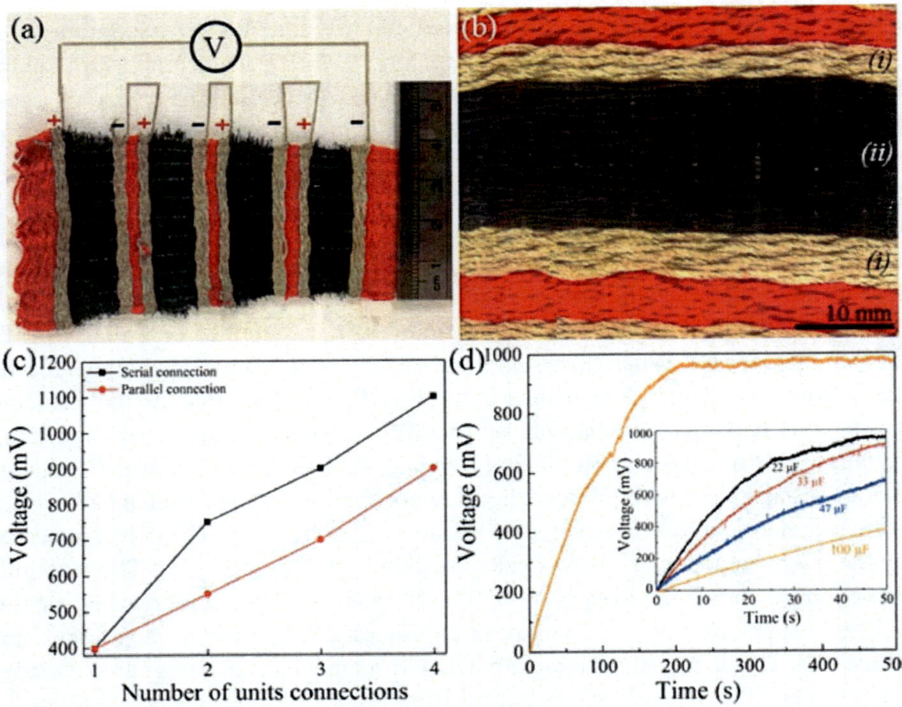

Fig. 5.17 **a** Photographs of woven PVDF/rGO fiber with four active units in serial connection, **b** magnified image of one active unit: (i) silver-coated nylon as electrodes and (ii) PVDF/rGO fibers, **c** output voltage for parallel connection with different active units connection, **d** charging 100 μF capacitor by the power generated from the four units connection in serial under periodic impact pressure

A bridge rectifier (four diodes of 1N5817) was placed in the circuit to feed a different capacitor under mechanical pressures in frequency of 1 Hz. Different capacitors (22, 33, 47, and 100 μF) charging performances were carried out upon mechanical deformation on the woven fabric with serial connection of units (Fig. 5.17d) and it is observed that the build up voltage increases exponentially to eventually reache a steady state. According to this woven fabric-based piezoelectric fiber, it possible to have piezoelectric fibers woven in different parts of a garment as the sensor or connected them in serial connection to have higher voltage to charge personal electronic devices. The PVDF/rGO piezoelectric fiber knitted in a circular knitting machine as the set up described in our previous work [16] (Fig. 5.18a).

The storage energy calculated for woven fabric with four active unit which connected in series was determined using capacitor potential energy Eq. (4.5) [16]. The energy storage based on the woven fabric in 100 μF capacitor was found to be 50 μJ. Based on the charging process of the 100 μF capacitor, the energy conversion efficiency of the woven energy generators is calculated as 26%.

The wearable energy generator could increase the voltage of the storage capacitor from 0 to 1 V in 200 s and the 100 μF capacitor was fully charged. The power output of the woven structure is 28 μW cm^{-3} under a periodic compression. This power is equal to the film power output of PVDF/(rGO–Ag) and film nanogenerator of rGO/PVDF-TrFE which recently reported [3, 11] with this significant difference that PVDF/rGO fibers have capability for mass production and have flexibility to be integrated to the textiles. To demonstrate potential applications of the developed circular knitted piezoelectric strain sensor for the detection of human and/or industrial activities, knitted wearable piezoelectric PVDF/rGO fibers were assembled in core–sheath structure as shown in Fig. 5.18b. The performance of the potential real-life applications of the developed wearable sensor under biomechanical pressure of periodic finger pressure and relaxation was demonstrated in Fig. 5.18c, d. In comparison with pure PVDF with the same structure, PVDF/rGO fibers shows 100% improvement in output voltage.

The structure of a circular knitted sensor filled inside and covered outside by silicon rubber to protect it from environmental condition (dust and humidity) for different applications is shown in Fig. 5.1e. The sensor output voltage was measured after different forces range from 0.5 to 10 N applied on 2 cm of sample by mechanical tester machine during compression test (Fig. 5.18e, f). As can be seen from Fig. 5.18f by increasing the force the output voltage increased and this sensor can be considered as a pressure sensor for wide range of applications (sensors in shoes, any part of clothes).

As depicted in Fig. 5.18f, the output voltage increases from 100 to 1200 mV in the pressure range of 8–160 kPa, and it demonstrates the sensitivity of 7.34 mV kPa^{-1} ($R^2 = 0.97$).

To verify potential health care application, the PVDF/rGO coils sensor with a length of 2 cm were fixed to a finger with an adjustable band as a prototype device. Silver coated nylon on both ends of the coil were included as electrodes. The coils could accurately identify the movements when the finger was bent and straightened (Fig. 5.19a). A clear result was obtained when coils generates different voltage signals at different bending angles. The higher the bending angle is, the higher the voltage is (Fig. 5.19b).

During the bending of the finger, the voltage increased and reached a peak value, whereas with the relaxation of the finger, the voltage decreased and returned to the initial position. The coil sensor can detect finger bending by generating voltage up to 730 mV. Here the frequency for the bending movement is approximately 0.5 Hz. The voltage response to the periodic motion of the finger validates that coil structure is stable in the monitoring process. The magnitude of the sensor signal (voltage and current) steadily increased almost linearly with the bending angle of the finger, thus representing the ability of the sensor to distinguish the degree of motion, as shown in Fig. 5.19c. The results confirmed that maximum peak-to-peak power of about 7 nW or 0.03 μW/cm^3 can be harvested from wearable piezoelectric coils by finger bending movements.

This coil structure is suitable to motion monitoring without disturbing the body's movements and was established to support personal recovery after a surgery. Also, the generated voltage can be transferred to the computer for hand motion detections in the VR environment for soft virtual reality glove systems. Moreover, the PVDF/rGO

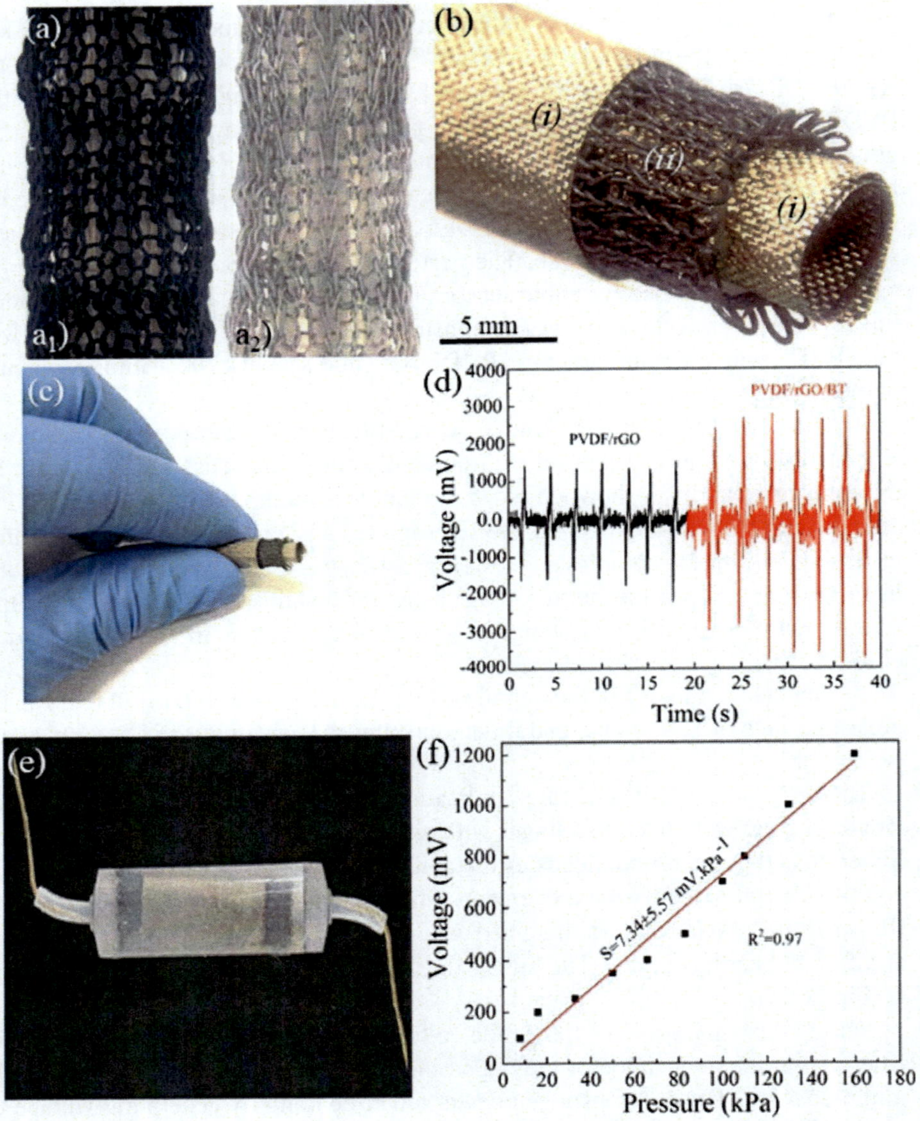

Fig. 5.18 **a** Circular knitted structure of: (a_1) PVDF/rGO and (a_2) PVDF/rGO/BT fiber, **b** optical photograph of the circular knitted piezogenerator: (i) conductive woven fabric as electrodes and (ii) knitted PVDF/rGO fiber as the middle layer, **c** assembled knitted piezo sensors with core–shell electrodes under hand pressure and **d** effect of pressure cycle on voltage output, **e** covered and filled knitted piezo sensors with silicone rubber as a pressure sensor and **f** voltage output of pressure sensor under different forces of mechanical tester

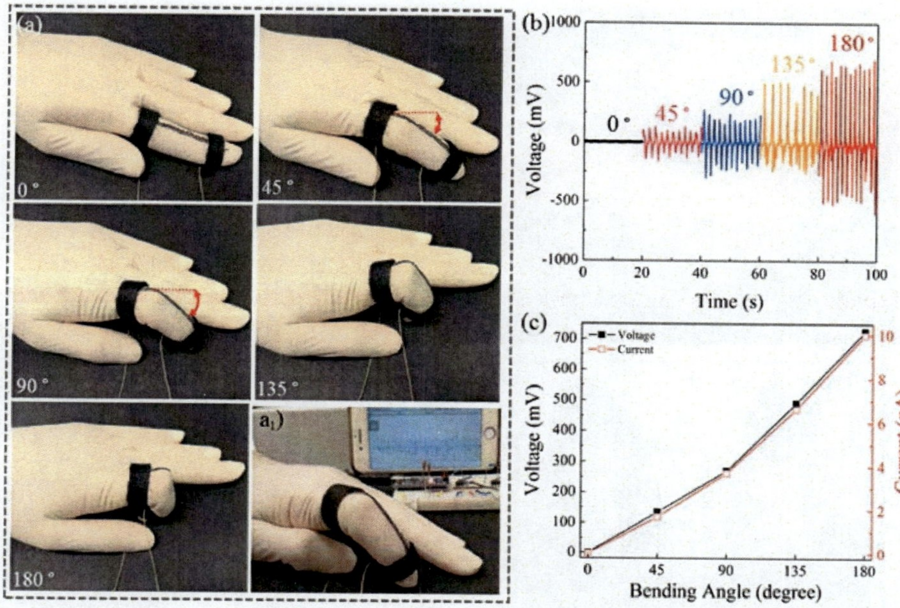

Fig. 5.19 **a** Practical demonstration of the PVDF/rGO coils as a human motion sensor for bending/straightening of finger in different angle, (a_1) experimental setup for demonstrating signals on phone, **b** the output voltage of PVDF/rGO coils at different bending angle in a bending and straightening cycle and **c** The relationships between different bending angle and output voltage and current of PVDF/rGO coils at frequency of 0.5 Hz [61]

coil also has a potential to be integrated in a stretchable 2D fabric used in close contact with the human skin or attached to textile as a decoration (Fig. 5.20a). PVDF/rGO coils set as warp along with elastomer fiber in a woven structure. Silver-coated nylon woven on both end as electrodes and acrylic fiber is weft direction made distance between two electrodes (Fig. 5.20a_1, a_2).

The poling voltage applied along the length of PVDF/rGO colis on woven silver coated nylons on both ends. The textile structure can be covered with soft and stretchable silicone material to provide protection from the environment including washing. The total weight of woven structure (2 × 5 cm) including the silver-coated nylon, coils and acrylic fibers is about 1 g. The energy-harvesting woven structure can be a potential strategy for building up a self-power system, which can be used to drive wearable electronics sustainably. Hence, using the PVDF/rGO coils, we attempted to simultaneously detect the muscle contraction and human arm motion in different angles during lifting two different weights (1 and 5 kg). The AnEx board can be attached to the fabric and linked to it by an external connector. The collected data can be transferred via Bluetooth to a laptop of phone for signal analysis (Fig. 5.20a). As can be seen from Fig. 5.20c, d the generated voltage for two weights of 1 and 5 kg have significant difference in the position of 90° but for maximum lifting of weight in 180° this difference decreased.

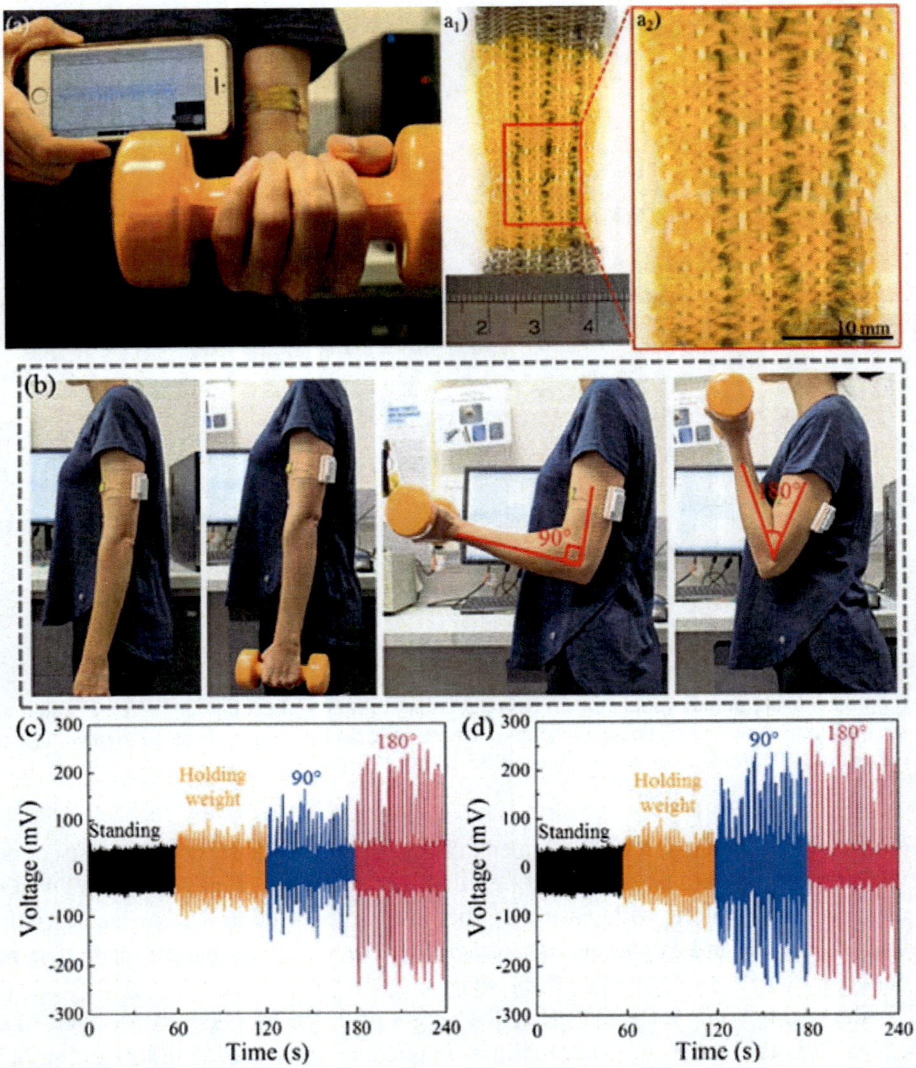

Fig. 5.20 Photograph of woven PVDF/rGO piezoelectric coils applied on arm, (a_1) the developed wearable and portable energy generator based on woven PVDF/rGO piezoelectric coils with the poling direction parallel to the length direction, (a_2) magnified image of woven structure with three PVDF/rGO coils in warp direction, **b** photograph of woven piezoelectric sensor based on PVDF/rGO coils during lifting weight, **c**, **d** the generated voltage associated with lifting weight of 1 and 5 kg in different positions

The generated voltage may change for different persons with high volume and strong muscles. Hence, this structure can be a lightweight and portable sensor and energy harvester to monitor body muscle during sport activity and diversified human motions. The development of elastomer woven structure is sensitive to all types of human motion and also capable of sensing pressure for various applications such as heart rate detection, muscle contraction, pressure monitoring, strain gauges, robots, etc., due to its ability to response to stretching. More importantly, developed processing method is scalable for the fabrication of industrial quantities of strain sensing and smart textiles.

5.5 Conclusion

In summary, the highly flexible nanocomposite piezoelectric fibers consisting of PVDF, BT and rGO were developed. The fibers were fabricated though a melt-spinning process. The fibers were strong enough to be coiled by twist insertion and then endure axial extension up to 100% strain. A high energy density of 80 mJ cm^{-2} was identified in the PVDF/rGO/BT nanocomposite owning to the rGO presence. These results demonstrated that PVDF-based graphene nanocomposites had better dielectric and conductivity properties. It is clear that rGO sheet are ordered and aligned in a uniform orientation, accounting for the improved significant enhancement in Young' modulus of the PVDF/rGO nanocomposite fiber. It was found that high-performance PVDF/rGO/BT piezofiber with 84% of the electroactive β-phase generated a maximum voltage output of 1.3 V and a power density 3 W/kg during longitudinal extension at 1 Hz. The coils could be easily fabricated through twist insertion. The piezoelectric response would be increased by increasing the original fiber diameter, length, and number. The coil structure from meltspun piezoelectric fibers is a novel method with ability of flexibility, lightweight and mass production capability. These coil structures have great potential application as motion detectors and self-powered biomedical and textile applications in daily life.

References

1. Habibur, R.M., Yaqoob, U., Muhammad, S., Uddin, A.S.M.I., Kim, H.C.: The effect of RGO on dielectric and energy harvesting properties of P(VDF-TrFE) matrix by optimizing electroactive β phase without traditional polling process. Mater. Chem. Phys. **215**, 46–55 (2018)
2. Si, S.K., Karan, S.K., Paria, S., Maitra, A., Das, A.K., Bera, R., Bera, A., Halder, L., Khatua, B.B.: A strategy to develop an efficient piezoelectric nanogenerator through ZTO assisted γ-phase nucleation of PVDF in ZTO/PVDF nanocomposite for harvesting bio-mechanical energy and energy storage application. Mater. Chem. Phys. **213**, 525–537 (2018)
3. Hu, X., Ding, Z., Fei, L., Xiang, Y., Lin, Y.: Wearable piezoelectric nanogenerators based on reduced graphene oxide and in situ polarization-enhanced PVDF-TrFE films. J. Mater. Sci. **54**(8), 6401–6409 (2019)
4. Cacciotti, I., Valentini, M., Raio, M., Nanni, F.: Design and development of advanced $BaTiO_3$/MWCNTs/PVDF multi-layered systems for microwave applications. Compos. Struct. **224**, 111075 (2019)
5. Sun, B., Li, X., Zhao, R., Ji, H., Qiu, J., Zhang, N., He, D., Wang, C.: Electrospun poly (vinylidene fluoride)-zinc oxide hierarchical composite fiber membrane as piezoelectric acoustoelectric nanogenerator. J. Mater. Sci. **54**(3), 2754–2762 (2019)
6. Ippili, S., Jella, V., Eom, J.-H., Kim, J., Hong, S., Choi, J.-S., Tran, V.-D., Van Hieu, N., Kim, Y.-J., Kim, H.-J.: An eco-friendly flexible piezoelectric energy harvester that delivers high

output performance is based on lead-free $MASnI_3$ films and $MASnI_3$-PVDF composite films. Nano Energy **57**, 911–923 (2019)
7. Su, Y.-F., Kotian, R.R., Lu, N.: Energy harvesting potential of bendable concrete using polymer based piezoelectric generator. Compos. B **153**, 124–129 (2018)
8. Javadi, A., Xiao, Y., Xu, W., Gong, S.: Chemically modified graphene/P(VDF-TrFE-CFE) electroactive polymer nanocomposites with superior electromechanical performance. J. Mater. Chem. **22**(3), 830–834 (2012)
9. Qi, F., Chen, N., Wang, Q.: Dielectric and piezoelectric properties in selective laser sintered polyamide11/$BaTiO_3$/CNT ternary nanocomposites. Mater. Des. **143**, 72–80 (2018)
10. Ansari, S., Giannelis, E.P.: Functionalized graphene sheet—poly(vinylidene fluoride) conductive nanocomposites. J. Polym. Sci. Part B Polym. Phys. **47**(9), 888–897 (2009)
11. Pusty, M., Sinha, L., Shirage, P.M.: A flexible self-poled piezoelectric nanogenerator based on a rGO–Ag/PVDF nanocomposite. New J. Chem. **43**(1), 284–294 (2019)
12. Hazarika, A., Deka, B.K., Jeong, C., Park, Y.-B., Park, H.W.: Biomechanical energy-harvesting wearable textile-based personal thermal management device containing epitaxially grown aligned Ag-tipped-$Ni_xCo_{1-x}Se$ nanowires/reduced graphene oxide. Adv. Funct. Mater. **29**(31), 1903144 (2019)
13. Karan, S.K., Mandal, D., Khatua, B.B.: Self-powered flexible Fe-doped RGO/PVDF nanocomposite: an excellent material for a piezoelectric energy harvester. Nanoscale **7**(24), 10655–10666 (2015)
14. Ataur Rahman, M., Lee, B.-C., Phan, D.-T., Chung, G.-S.: Fabrication and characterization of highly efficient flexible energy harvesters using PVDF–graphene nanocomposites. Smart Mater. Struct. **22**(8), 085017 (2013)
15. Alamusi, Xue, J., Wu, L., Hu, N., Qiu, J., Chang, C., Atobe, S., Fukunaga, H., Watanabe, T., Liu, Y., Ning, H., Li, J., Li, Y., Zhao, Y.: Evaluation of piezoelectric property of reduced graphene oxide (rGO)–poly(vinylidene fluoride) nanocomposites. Nanoscale **4**(22), 7250–7255 (2012)
16. Mokhtari, F., Spinks, G.M., Fay, C., Cheng, Z., Raad, R., Xi, J., Foroughi, J.: Wearable electronic textiles from nanostructured piezoelectric fibers. Adv. Mater. Technol. **n/a**(n/a), 1900900
17. Mokhtari, F., Foroughi, J., Zheng, T., Cheng, Z., Spinks, G.M.: Triaxial braided piezo fiber energy harvesters for self-powered wearable technologies. J. Mater. Chem. A **7**(14), 8245–8257 (2019)
18. Huang, L., Lu, C., Wang, F., Wang, L.: Preparation of PVDF/graphene ferroelectric composite films by in situ reduction with hydrobromic acids and their properties. RSC Adv. **4**(85), 45220–45229 (2014)
19. Yaqoob, U., Uddin, A.S.M.I., Chung, G.-S.: The effect of reduced graphene oxide on the dielectric and ferroelectric properties of PVDF–$BaTiO_3$ nanocomposites. RSC Adv. **6**(36), 30747–30754 (2016)
20. Ghosh, S.K., Alam, M.M., Mandal, D.: The in situ formation of platinum nanoparticles and their catalytic role in electroactive phase formation in poly(vinylidene fluoride): a simple preparation of multifunctional poly(vinylidene fluoride) films doped with platinum nanoparticles. RSC Adv. **4**(79), 41886–41894 (2014)
21. Sabira, K., Saheeda, P., Divyasree, M.C., Jayalekshmi, S.: Impressive nonlinear optical response exhibited by poly(vinylidene fluoride) (PVDF)/reduced graphene oxide (RGO) nanocomposite films. Opt. Laser Technol. **97**, 77–83 (2017)
22. Shi, K., Sun, B., Huang, X., Jiang, P.: Synergistic effect of graphene nanosheet and $BaTiO_3$ nanoparticles on performance enhancement of electrospun PVDF nanofiber mat for flexible piezoelectric nanogenerators. Nano Energy **52**, 153–162 (2018)
23. Zhang, C., Zhou, T., Tang, W., Han, C., Zhang, L., Wang, Z.L.: Rotating-disk-based direct-current triboelectric nanogenerator. Adv. Energy Mater. **4**(9), 1301798 (2014)
24. Karan, S.K., Bera, R., Paria, S., Das, A.K., Maiti, S., Maitra, A., Khatua, B.B.: An approach to design highly durable piezoelectric nanogenerator based on self-poled PVDF/AlO-rGO flexible nanocomposite with high power density and energy conversion efficiency. Adv. Energy Mater. **6**(20), 1601016 (2016)

25. Kumar, C., Gaur, A., Rai, S.K., Maiti, P.: Piezo devices using poly(vinylidene fluoride)/reduced graphene oxide hybrid for energy harvesting. Nano-Struct. Nano-Obj. **12**, 174–181 (2017)
26. Ponnamma, D., Goutham, S., Sadasivuni, K.K., Rao, K.V., Cabibihan, J.J., Al-Maadeed, M.A.A.: Controlling the sensing performance of rGO filled PVDF nanocomposite with the addition of secondary nanofillers. Synth. Met. **243**, 34–43 (2018)
27. Zhu, P., Zhu, J., Zang, J., Chen, C., Lu, Y., Jiang, M., Yan, C., Dirican, M., Kalai Selvan, R., Zhang, X.: A novel bi-functional double-layer rGO–PVDF/PVDF composite nanofiber membrane separator with enhanced thermal stability and effective polysulfide inhibition for high-performance lithium–sulfur batteries. J. Mater. Chem. A **5**(29), 15096–15104 (2017)
28. El Achaby, M., Arrakhiz, F.Z., Vaudreuil, S., Essassi, E.M., Qaiss, A.: Piezoelectric β-polymorph formation and properties enhancement in graphene oxide—PVDF nanocomposite films. Appl. Surf. Sci. **258**(19), 7668–7677 (2012)
29. Sieradzka, M., Fryczkowski, R., Biniaś, D., Biniaś, W., Janicki, J.: A facile approach to obtaining PVDF/graphene fibers and the effect of nanoadditive on the structure and properties of nanocomposites. Polym. Testing **81**, 106229 (2020)
30. Al-Saygh, A., Ponnamma, D., AlMaadeed, M.A., Vijayan, P.P., Karim, A., Hassan, M.K.: Flexible pressure sensor based on PVDF nanocomposites containing reduced graphene oxide-titania hybrid nanolayers. Polymers **9**(2), 33 (2017)
31. Lund, A., Rundqvist, K., Nilsson, E., Yu, L., Hagström, B., Müller, C.: Energy harvesting textiles for a rainy day: woven piezoelectrics based on melt-spun PVDF microfibres with a conducting core. NPJ Flex. Electron. **2**(1), 9 (2018)
32. Li, B., Zhang, F., Guan, S., Zheng, J., Xu, C.: Wearable piezoelectric device assembled by one-step continuous electrospinning. J. Mater. Chem. C **4**(29), 6988–6995 (2016)
33. Ponnamma, D., Sadasivuni, K.K., Strankowski, M., Guo, Q., Thomas, S.: Synergistic effect of multi walled carbon nanotubes and reduced graphene oxides in natural rubber for sensing application. Soft Matter **9**(43), 10343–10353 (2013)
34. Olad, A., Nosrati, R.: Preparation and corrosion resistance of nanostructured PVC/ZnO–polyaniline hybrid coating. Prog. Org. Coat. **76**(1), 113–118 (2013)
35. Bindu Sharmila, T.K., Antony, J.V., Jayakrishnan, M.P., Sabura Beegum, P.M., Thachil, E.T.: Mechanical, thermal and dielectric properties of hybrid composites of epoxy and reduced graphene oxide/iron oxide. Mater. Des. **90**, 66–75 (2016)
36. Song, R., Yang, D., He, L.: Effect of surface modification of nanosilica on crystallization, thermal and mechanical properties of poly(vinylidene fluoride). J. Mater. Sci. **42**(20), 8408–8417 (2007)
37. Yang, J.-H., Xie, X., He, Z.-Z., Lu, Y., Qi, X.-D., Wang, Y.: Graphene oxide-tailored dispersion of hybrid barium titanate@polypyrrole particles and the dielectric composites. Chem. Eng. J. **355**, 137–149 (2019)
38. Bera, M., Saha, U., Bhardwaj, A., Maji, P.K.: Reduced graphene oxide (RGO)-induced compatibilization and reinforcement of poly(vinylidene fluoride) (PVDF)–thermoplastic polyurethane (TPU) binary polymer blend. J. Appl. Polym. Sci. **136**(5), 47010 (2019)
39. Li, X., Lu, H., Li, J., Dong, G.: Preparation and lubricating properties of poly(vinylidene-fluoride) particles wrapped by reduced graphene oxide. Tribol. Int. **127**, 351–360 (2018)
40. Tong, J., Huang, H.-X., Wu, M.: Simultaneously facilitating dispersion and thermal reduction of graphene oxide to enhance thermal conductivity of poly(vinylidene fluoride)/graphene nanocomposites by water in continuous extrusion. Chem. Eng. J. **348**, 693–703 (2018)
41. Jaleh, B., Jabbari, A.: Evaluation of reduced graphene oxide/ZnO effect on properties of PVDF nanocomposite films. Appl. Surf. Sci. **320**, 339–347 (2014)
42. Zhu, W., Ma, J., Nan, X., Osei Lartey, P., Yang, Y.: Study on dispersion of reduced graphene oxide on physical performance of Polyvinylidene fluoride composites by Hansen solubility parameters. Colloid Polym. Sci. **297**(2), 213–224 (2019)
43. Maity, N., Mandal, A., Nandi, A.K.: Synergistic interfacial effect of polymer stabilized graphene via non-covalent functionalization in poly(vinylidene fluoride) matrix yielding superior mechanical and electronic properties. Polymer **88**, 79–93 (2016)

44. Maity, N., Mandal, A., Nandi, A.K.: Interface engineering of ionic liquid integrated graphene in poly(vinylidene fluoride) matrix yielding magnificent improvement in mechanical, electrical and dielectric properties. Polymer **65**, 154–167 (2015)
45. Mandal, A., Nandi, A.K.: Ionic liquid integrated multiwalled carbon nanotube in a poly(vinylidene fluoride) matrix: formation of a piezoelectric β-polymorph with significant reinforcement and conductivity improvement. ACS Appl. Mater. Interfaces **5**(3), 747–760 (2013)
46. Cho, S., Kim, M., Lee, J.S., Jang, J.: Polypropylene/polyaniline nanofiber/reduced graphene oxide nanocomposite with enhanced electrical, dielectric, and ferroelectric properties for a high energy density capacitor. ACS Appl. Mater. Interfaces **7**(40), 22301–22314 (2015)
47. Hou, Y., Deng, Y., Wang, Y., Gao, H.: Uniform distribution of low content $BaTiO_3$ nanoparticles in poly(vinylidene fluoride) nanocomposite: toward high dielectric breakdown strength and energy storage density. RSC Adv. **5**(88), 72090–72098 (2015)
48. Zhang, D., Zhou, X., Roscow, J., Zhou, K., Wang, L., Luo, H., Bowen, C.R.: Significantly enhanced energy storage density by modulating the aspect ratio of $BaTiO_3$ nanofibers. Sci. Rep. **7**(1), 45179 (2017)
49. Ataur Rahman, M., Chung, G.-S.: Synthesis of PVDF-graphene nanocomposites and their properties. J. Alloy. Compd. **581**, 724–730 (2013)
50. Zhu, Y., Murali, S., Cai, W., Li, X., Suk, J.W., Potts, J.R., Ruoff, R.S.: Graphene and graphene oxide: synthesis, properties, and applications. Adv. Mater. **22**(35), 3906–3924 (2010)
51. He, F., Lau, S., Chan, H.L., Fan, J.: High dielectric permittivity and low percolation threshold in nanocomposites based on poly(vinylidene fluoride) and exfoliated graphite nanoplates. Adv. Mater. **21**(6), 710–715 (2009)
52. Uyor, U.O., Popoola, A.P., Popoola, O., Aigbodion, V.S.: Energy storage and loss capacity of graphene-reinforced poly(vinylidene fluoride) nanocomposites from electrical and dielectric properties perspective: a review. Adv. Polym. Technol. **37**(8), 2838–2858 (2018)
53. Mallick, S., Ahmad, Z., Qadir, K.W., Rehman, A., Shakoor, R.A., Touati, F., Al-Muhtaseb, S.A.: Effect of $BaTiO_3$ on the sensing properties of PVDF composite-based capacitive humidity sensors. Ceram. Int. **46**(3), 2949–2953 (2020)
54. Mohammed, M.I., Fouad, S.S., Mehta, N.: Dielectric relaxation and thermally activated A.C. conduction in (PVDF)/(rGO) nano-composites: role of rGO over different fillers. J. Mater. Sci. Mater. Electron. **29**(21), 18271–18281 (2018)
55. Yaqoob, U., Chung, G.-S.: Effect of surface treated MWCNTs and $BaTiO_3$ nanoparticles on the dielectric properties of a P(VDF-TrFE) matrix. J. Alloy. Compd. **695**, 1231–1236 (2017)
56. Haines, C.S., Lima, M.D., Li, N., Spinks, G.M., Foroughi, J., Madden, J.D.W., Kim, S.H., Fang, S., Jung de Andrade, M., Göktepe, F., Göktepe, Ö., Mirvakili, S.M., Naficy, S., Lepró, X., Oh, J., Kozlov, M.E., Kim, S.J., Xu, X., Swedlove, B.J., Wallace, G.G., Baughman, R.H.: Artificial muscles from fishing line and sewing thread. Science **343**(6173), 868–872 (2014)
57. Kim, S.H., Haines, C.S., Li, N., Kim, K.J., Mun, T.J., Choi, C., Di, J., Oh, Y.J., Oviedo, J.P., Bykova, J., Fang, S., Jiang, N., Liu, Z., Wang, R., Kumar, P., Qiao, R., Priya, S., Cho, K., Kim, M., Lucas, M.S., Drummy, L.F., Maruyama, B., Lee, D.Y., Lepró, X., Gao, E., Albarq, D., Ovalle-Robles, R., Kim, S.J., Baughman, R.H.: Harvesting electrical energy from carbon nanotube yarn twist. Science **357**(6353), 773–778 (2017)
58. Sim, H.J., Choi, C., Lee, C.J., Kim, Y.T., Spinks, G.M., Lima, M.D., Baughman, R.H., Kim, S.J.: Flexible, stretchable and weavable piezoelectric fiber. Adv. Eng. Mater. **17**(9), 1270–1275 (2015)
59. Kim, S.H., Sim, H.J., Hyeon, J.S., Suh, D., Spinks, G.M., Baughman, R.H., Kim, S.J.: Harvesting electrical energy from torsional thermal actuation driven by natural convection. Sci. Rep. **8**(1), 8712 (2018)
60. Chang, C., Tran, V.H., Wang, J., Fuh, Y.-K., Lin, L.: Direct-write piezoelectric polymeric nanogenerator with high energy conversion efficiency. Nano Lett. **10**(2), 726–731 (2010)
61. Mokhtari, F., Spinks, G.M., Sayyar, S., Cheng, Z., Ruhparwar, A., Foroughi, J.: Highly stretchable self-powered wearable electrical energy generator and sensors. Adv. Mater. Technol. **6**(2), 2000841 (2021)

Chapter 6
Conclusions and Future Works

6.1 General Conclusion

The main objective of this thesis was developing piezoelectric fibers and textiles for sensing and energy generation applications including the real-time monitoring of human movements and micro-power-generation in E-Textile industries. To reduce cost and bring real impact to society, large-scale production of energy harvesting smart textiles is an important factor. Therefore, poly(vinylidene fluoride) (PVDF) as a piezoelectric polymer was fabricated in the form of fibers using melt spinning that demonstrated the ability for mass production and fabrication of flexible microfibres. These piezofibers have the ability to generate voltage after any kind of deformation such as bending compression, impact and stretching. Hence, these fibers were considered suitable for integration into textile structures that were to be investigated as self-powered wireless sensors, structural and human health monitoring systems, and cheaply harvesting energy from human movements.

The experimental studies in Chap. 3 showed that the melt spinning parameters greatly influence the formation of β phase crystallinity, which is necessary for the piezoelectric activity in the fibers. In order to enhance the piezoelectric performance of fibers post treatment procedures such as cold drawing and poling process were optimised. Cold drawing is a vital part of any melt spinning process and it was found that the optimised temperature and draw ratio were 80 °C and 4, respectively, to provide the maximum β phase fraction. The poling process was performed using two different technique of corona and contact with an electric field strength of 150 MV/m. In both methods, the most important parameter for efficient polarisation was a high applied voltage. Piezo response Force Microscopy (PFM) measurements identified a variation in the piezoelectric response between different points in the fibers if they have a length more than 1 cm during the poling process. This result suggests that the nanoscale piezoelectricity depends on variations on the local properties including polarization and degree of crystallinity.

The first textile device was constructed with the poled piezofibers by a multi-step braiding process to construct a novel triaxial braided piezofiber. This design allowed

F. Mokhtari, *Self-Powered Smart Fabrics for Wearable Technologies*, Springer Theses,
https://doi.org/10.1007/978-3-031-06481-4_6

the electrodes to be embedded as a core-shell structure and braided piezoelectric fibers placed in between the inner and outer electrodes. The major difference in the triaxial braided piezofiber developed as part of this study compared to conventional piezo fibers was that it could be easily fabricated with unlimited lengths. The developed triaxial braided piezofiber used silver coated nylon electrodes which bring more flexibility and durability for the structure. The final fibres were shown to be highly sensitive to motion and impact testing. The PVDF triaxial braided fiber unexpectedly exhibited a power density of 29.62 $\mu W\ cm^{-3}$ which is ~1559% higher than previously reported (1.78 $\mu W\ cm^{-3}$) for piezoelectric textiles. The comparison between the power outputs of the developed triaxial braided piezoelectric energy harvester and pervious piezoelectric PVDF energy generators is illustrated in Fig. 6.1.

The next phase of this thesis considered the use of filler materials in the forms of nanoparticle added to the polymer matrix as a potential means for improving the piezoelectric properties of the melt spun piezofibers. Piezoelectric ceramic particles play an important role in enhancing the efficiency of piezoelectric fiber composites. Therefore, barium titanate (BT) nanoparticles as the inorganic piezoelectric materials was chosen to be added to the PVDF matrix. The BT ceramic particles have high intrinsic piezoelectric performance, high dielectric constant ($\varepsilon \sim 2000$), low cost, natural abundance, and environmental friendliness. Since a number of processing parameters are required to be optimised and controlled in order to form composite fibres and also because of the limited number of existing studies in the area of PVDF/BT melt spun fibers, it was necessary to initially investigate the optimum conditions for nanocomposite fiber fabrication. This study was carried out and the

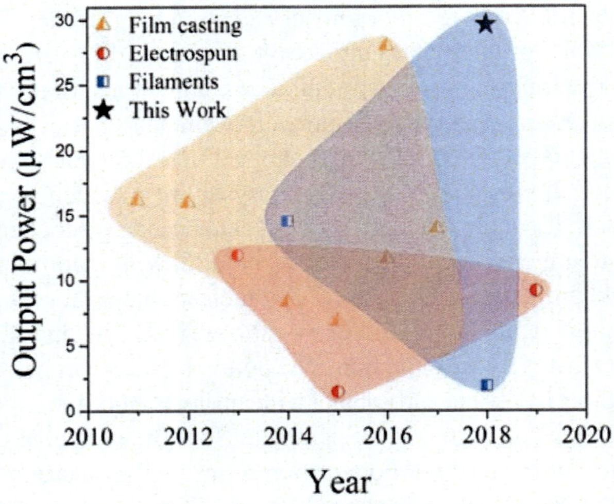

Fig. 6.1 Comparison of the power density for the present triaxial braided energy harvesting generator and previously reported energy generators based on pure piezoelectric PVDF films and fibers. The maximum power density for the present triaxial braided energy harvester is comparable to or higher than that in previous studies; however, its sensitivity (output voltage the applied force) significantly exceeds that in previous studies (▲ [1–7], ◐ [8–10], and ◧ [11, 12]).

results are presented in Chap. 4. Selection of appropriate filler content, as well as the fabrication and post treatment parameters were critical factors that were successfully addressed in these studies. The results showed that the fibres which contained 10% BT yielded the optimum mechanical results with Young's modulus of ~891 MPa and ultimate stress of ~168 MPa compared to fibres produced with other components. The β-phase content of the PVDF/BT nanocomposite fibers first increased and then decreased as the amount of BT added to the PVDF increased. The amount of β-phase decreased as a result of increasing BT percentage for more than 10 wt%. The incorporation of BT nanoparticles was found to increase the dielectric constant of fiber due to the large dielectric constant of BT. The dielectric constant gradually decreased when measured at higher frequencies.

The hybrid BT-PVDF piezoelectric fibers were used to form a variety of textile architectures including circular knitting, weaving and braiding structures. In all these structures the conductive fiber and fabrics were used to avoid the negative impact on textile softness, light weight, breathability, fit, and tactile comfort. The energy storage based on the as-fabricated textile energy harvesters in 10 μF capacitor was found to be 36.45 μJ for circular knitting structure which was higher than other structures. The energy conversion efficiency of the woven energy generators was 40% which established the viability of them in real-life applications. The potential applications of the developed textile structures as a strain sensor for the detection of human and/or industrial activities and hydraulic/pneumatic pressure sensors with tuneable sensitivity were evaluated. The sensors were coupled with a portable wireless device to allow data to be transferred to a laptop for signal analysis. The results revealed that textile structures have the stability performance for long time loading measurements. In Table 6.1 the woven PVDF/BT structure in case of output performance and energy conversion efficiency compared with previous piezoelectric generators.

Further potential improvements to the piezoelectric performance of PVDF fibers were considered by combining BT and rGO as nanoparticle additives to the PVDF matrix. These studies are described in Chap. 5. The PVDF composite fibers were used to make different structures including coil, woven and knitted textiles. The performance of different composites of PVDF, PVDF/BT, PVDF/rGO and PVDF/rGO/BT in all these structures were compared. Among all the samples, the PVDF/rGO/BT nanocomposite coil showed a maximum voltage output of 1240 mV at an applied force of 1 N for 100% strain when pure PVDF coil generated a maximum voltage output of 300 mV while PVDF/rGO and PVDF/BT samples showed 400 mV and 700 m V maximum output voltages, respectively. The PVDF/rGO fiber was fabricated in the plain weave structure with silver coated nylon as electrodes. With four active sections of the woven structure connected in series it could fully charge a 100 μF capacitor up to 1 V with the energy conversion efficiency of 26%. The PVDF/rGO fibers in a circular knitted structure demonstrated the sensitivity is 7.34 mV kPa^{-1} during compression loading. In a portable and adjustable setup the PVDF/rGO coil structure could accurately identify finger bending up to 180° and generated a voltage of 730 mV. The results confirmed that maximum peak-to-peak power of about 7 nW or 0.03 $\mu W/cm^3$ can be harvested from wearable piezoelectric coils by finger bending movements. In another demonstration, the PVDF/rGO coils with silver-coated nylon

Table 6.1 Comparison of the output performance and the energy conversion efficiency of PVDF and PVDF composites energy harvesters

Active material	Active area	Output voltage (V)	Output power	Energy conversion efficiency (%)	References
PVDF	5 cm × 5 cm	0.03	–	21.8	[13]
PMLG/PVDF	5 mm × 5 mm	0.08	0.0025 μW cm^{-2}	3.3	[14]
PVDF-TrFE	–	3	16.5 nW	11	[15]
PVDF/AlO-rGO	–	6.1	27.97 μW cm^{-3}	12.47	[7]
Gr-BT/PVDF	2.5 cm × 2.5 cm	11	0.656 μW cm^{-2}	1.69	[16]
Yb^{3+}/PVDF	1.6 cm × 3.2 cm	10	1 μW cm^{-2}	2.4	[17]
PVDF/ZnS-NRs	8 cm × 8 cm	6	0.15 μW cm^{-2}	58	[18]
PVDF/TiO_2	–	1.8	–	70	[19]
knitted PVDF/BT_{10}	1 cm × 2 cm	1	87 μW cm^{-3}	39.5	This work

were woven together and use to monitor muscle contraction and human arm motion in different angles during lifting weights. This application can be useful to monitor muscle performance of athletes during their exercise or in the recovery of muscle injury which is the most common complaints in orthopedic practice, occurring both among athletes and among non-athletes. The more powerful muscle contractions generated more voltage and this signal can evaluate the muscle health. Therefore, it can be concluded the PVDF-based textiles exhibits broad prospects in the development of the human-based self-powered area, and it is significant in electronic system devices and medical sensing applications. Table 6.2 summarize the performance of all devices developed in this thesis.

6.2 Suggestions for Future Work

The main aim of this thesis was to investigate the properties of piezoelectric fiber at the microscale which contain different fillers to enhance their output performance. This goal was successfully achieved from fundamental study to fibers integration in different textile structures. However, like any other research, there are many promising areas still to be continued in future research which could not be address in this thesis. So, here I will suggest the areas of research that should be further explored, as regards the results presented in this thesis.

Table 6.2 Comparison of the performance of all PVDF base developed devices

Materials	Type	Active area size and weight	Voltage (mV)	Power density ($\mu W\ cm^{-3}$)	Excitation method
PVDF	Braid	2 cm × 2.5 mm, 0.0138 g cm^{-1}	380	29.62	Falling ball
PVDF/BT	Woven	3.14 cm^2, 0.057 g cm^{-2}	1000	36.2	Periodic compression
PVDF/BT	Braid	2 cm × 2.5 mm, 0.08 g cm^{-1}	1500	38.8	Periodic compression
PVDF/BT	Knit	2 cm × 1 cm, 0.034 g cm^{-1}	4000	87	Periodic compression
PVDF/rGO	Coil	2 cm, 0.001 g cm^{-1}	700	0.03	Finger bending
PVDF/rGO/BT	Coil	1 cm, 0.0015 g cm^{-1}	1200	0.42	100% strain
PVDF/rGO	Woven	4 cm × 10 mm, 3 g	1000	28	Periodic pressure
PVDF/rGO	Knit	2 cm × 1 cm, 0.028 g cm^{-1}	1200	57	Compression

- The output currents of the piezoelectric fibers were in the range of nA. To increase the output current, different fillers are needed to be used for further investigate of composite fibers. In this thesis the fillers of rGO and BT and mixture of them were evaluated other functional materials, hybridization mechanism, conductive filler and piezo ceramic particles need to be explored. Also use of PVDF co-polymers is highly recommended.
- Deeper study about polarization uniformity and depolarization in piezoelectric fibers contain different fillers needs to be established by the PFM method. This method provides imaging of ferroelectric domains on a sub-micrometer scale during switching events. This information can help elucidate the links between microstructure and piezoelectric parameters to help optimise the structure for enhance piezoelectric performance.
- Selecting appropriate electrodes based on the final application condition is another important factor. These electrodes must collect the charge from generator (piezo) part and transfer it to the target user. Therefore, these electrodes should have a suitable interface with the generator and also have low internal resistance. Embedding electrodes in an appropriate contact with active area of piezo generator is another factor which should be considered.
- The processing and post processing conditions for fiber fabrication during melt spinning method including feeding and cooling rate, take up speed, drawing ratio and temperature, poling time, voltage and temperature for different fillers with different concentration need to be discussed more. Also, it is suggested to evaluate the effect of environment conditions such as temperature and humidity on piezoelectric response of fibers.

- Effective packaging of the device in case of integrating of materials and electrodes in a textile-based device is necessary to have a durable product. However, the packaging also needs to ensure breathability, comfort, and wearability. Also wearable generators are exposed to continuous movements and complicated deformation which brings noise to the system. The appropriate packaging system should somehow filter noise.
- The excitation frequency range and amount of strain on output performance of piezo generator with different fillers content need to be evaluated.
- There is an opportunity to develop improved energy harvesting garments by combining piezoelectric, triboelectric and electrostatic effect to improve the performance of the generator.
- There should be a capable system to harvest the power generated by the device and guide this energy to be used for final application. This system can be a lithium ion battery with low self-discharge and high energy density. Hence, more research is required to develop an effective energy harvesting circuit for charging a battery.
- It is recommended to model piezoelectric fibers in simulation software such as ANSYS, Solid Works or ABAQUS. In the modelling environment it is possible to predict the output voltage and efficiency of generator in different deformation forms including bending, stretching and twisting. Also, the effect of structure shape and dimension can be evaluated by simulation technique.
- The piezoelectric fibres applications are in various areas such as medicine, sport, industry sensors, constructions which need to be considered and discussed with professionals from these fields.

References

1. Vatansever, D., Hadimani, R., Shah, T., Siores, E.: An investigation of energy harvesting from renewable sources with PVDF and PZT. Smart Mater. Struct. **20**(5), 055019 (2011)
2. Lee, M., Chen, C.-Y., Wang, S., Cha, S.N., Park, Y.J., Kim, J.M., Chou, L.-J., Wang, Z.L.: A hybrid piezoelectric structure for wearable nanogenerators. Adv. Mater. **24**(13), 1759–1764 (2012)
3. Chen, D., Sharma, T., Zhang, J.X.J.: Mesoporous surface control of PVDF thin films for enhanced piezoelectric energy generation. Sens. Actuators A **216**, 196 (2014)
4. Kumar, C., Gaur, A., Rai, S.K., Maiti, P.: Piezo devices using poly(vinylidene fluoride)/reduced graphene oxide hybrid for energy harvesting. Nano-Struct. Nano-Objects **12**, 174 (2017)
5. Yoon, S.-J., Arakawa, K., Uchino, M.: Development of an energy harvesting damper using PVDF film. Int. J. Energy Res. **39**(11), 1545–1553 (2015)
6. Emamian, S., Chlaihawi, A.A., Narakathu, B.B., Bazuin, B.J., Atashbar, M.Z.: A piezoelectric based vibration energy harvester fabricated using screen printing technique. In: 2016 IEEE Sensors, p. 1, 30 Oct 2016–3 Nov 2016 (2016)
7. Karan, S.K., Bera, R., Paria, S., Das, A.K., Maiti, S., Maitra, A., Khatua, B.B.: An approach to design highly durable piezoelectric nanogenerator based on self-poled PVDF/AlO-rGO flexible nanocomposite with high power density and energy conversion efficiency. Adv. Energy Mater. **6**(20), 1601016 (2016)
8. Zeng, W., Tao, X.-M., Chen, S., Shang, S., Chan, H.L.W., Choy, S.H.: Highly durable all-fiber nanogenerator for mechanical energy harvesting. Energy Environ. Sci. **6**(9), 2631–2638 (2013)

9. Fuh, Y.-K., Chen, P.-C., Ho, H.-C., Huang, Z.-M., Li, S.-C.: All-direction energy harvester based on nano/micro fibers as flexible and stretchable sensors for human motion detection. RSC Adv. **5**(83), 67787 (2015)
10. Hwang, Y.J., Choi, S., Kim, H.S.: Highly flexible all-nonwoven piezoelectric generators based on electrospun poly(vinylidene fluoride). Sens. Actuators A **300**, 111672 (2019)
11. Lund, A., Rundqvist, K., Nilsson, E., Yu, L., Hagström, B., Müller, C.: Energy harvesting textiles for a rainy day: woven piezoelectrics based on melt-spun PVDF microfibres with a conducting core. NPJ Flex. Electron. **2**(1), 9 (2018)
12. Soin, N., Shah, T.H., Anand, S.C., Geng, J., Pornwannachai, W., Mandal, P., Reid, D., Sharma, S., Hadimani, R.L., Bayramol, D.V.: Novel "3-D spacer" all fibre piezoelectric textiles for energy harvesting applications. Energy Environ. Sci. **7**(5), 1670–1679 (2014)
13. Chang, C., Tran, V.H., Wang, J., Fuh, Y.-K., Lin, L.: Direct-write piezoelectric polymeric nanogenerator with high energy conversion efficiency. Nano Lett. **10**(2), 726–731 (2010)
14. Pan, C.-T., Yen, C.-K., Wu, H.-C., Lin, L., Lu, Y.-S., Huang, J.C.-C., Kuo, S.-W.: Significant piezoelectric and energy harvesting enhancement of poly(vinylidene fluoride)/polypeptide fiber composites prepared through near-field electrospinning. J. Mater. Chem. A **3**(13), 6835–6843 (2015)
15. Jing, Q., Kar-Narayan, S.: Nanostructured polymer-based piezoelectric and triboelectric materials and devices for energy harvesting applications. J. Phys. D Appl. Phys. **51**(30), 303001 (2018)
16. Shi, K., Sun, B., Huang, X., Jiang, P.: Synergistic effect of graphene nanosheet and $BaTiO_3$ nanoparticles on performance enhancement of electrospun PVDF nanofiber mat for flexible piezoelectric nanogenerators. Nano Energy **52**, 153–162 (2018)
17. Ghosh, S.K., Biswas, A., Sen, S., Das, C., Henkel, K., Schmeisser, D., Mandal, D.: Yb^{3+} assisted self-polarized PVDF based ferroelectretic nanogenerator: a facile strategy of highly efficient mechanical energy harvester fabrication. Nano Energy **30**, 621–629 (2016)
18. Sultana, A., Alam, M.M., Ghosh, S.K., Middya, T.R., Mandal, D.: Energy harvesting and self-powered microphone application on multifunctional inorganic-organic hybrid nanogenerator. Energy **166**, 963–971 (2019)
19. Alam, M.M., Ghosh, S.K., Sarkar, D., Sen, S., Mandal, D.: Improved dielectric constant and breakdown strength of γ-phase dominant super toughened polyvinylidene fluoride/TiO_2 nanocomposite film: an excellent material for energy storage applications and piezoelectric throughput. Nanotechnology **28**(1), 015503 (2016)